EVIDENCE AND METHOD

EVIDENCE AND METHOD

Scientific Strategies of Isaac Newton
and James Clerk Maxwell

Peter Achinstein

OXFORD
UNIVERSITY PRESS

Oxford University Press is a department of the University of Oxford.
It furthers the University's objective of excellence in research,
scholarship, and education by publishing worldwide.

Oxford New York
Auckland Cape Town Dar es Salaam Hong Kong Karachi
Kuala Lumpur Madrid Melbourne Mexico City Nairobi
New Delhi Shanghai Taipei Toronto

With offices in
Argentina Austria Brazil Chile Czech Republic France Greece
Guatemala Hungary Italy Japan Poland Portugal Singapore
South Korea Switzerland Thailand Turkey Ukraine Vietnam

Oxford is a registered trade mark of Oxford University Press
in the UK and certain other countries.

Published in the United States of America by
Oxford University Press
198 Madison Avenue, New York, NY 10016

© Oxford University Press 2013

All rights reserved. No part of this publication may be reproduced,
stored in a retrieval system, or transmitted, in any form or by any means, without the
prior permission in writing of Oxford University Press, or as expressly permitted by law,
by license, or under terms agreed with the appropriate reproduction rights organization.
Inquiries concerning reproduction outside the scope of the above should be sent
to the Rights Department, Oxford University Press, at the address above.

You must not circulate this work in any other form and
you must impose the same condition on any acquirer.

Library of Congress Cataloging-in-Publication Data
Achinstein, Peter.
Evidence and method : scientific strategies of Isaac Newton and
James Clerk Maxwell/Peter Achinstein.
pages cm
ISBN 978-0-19-992185-0 (alk. paper)
1. Science—Methodology—History—18th century. 2. Science—Methodology—History—
19th century. 3. Verification (Empiricism) 4. Newton, Isaac, 1642–1727.
5. Maxwell, James Clerk, 1831–1879. I. Title.
Q174.8.A238 2013
507.2—dc23   2012036610

In memory of Carl G. ("Peter") Hempel, Exemplar

CONTENTS

Preface	ix
Acknowledgments	xv
1. A Problem About Evidence	3
2. Newton's Rules	43
3. Newtonian Extensions, a Rival, Justifying Induction, and Evidence	84
4. What to Do If You Cannot Establish a Theory: Maxwell's Three Methods	127
Index	175

PREFACE

During the 1953–54 academic year, in my sophomore year at Harvard, Carl G. Hempel was a visiting professor, replacing W. V. Quine, who was on leave. He taught two courses that I took, one in logic, one in probability. Another course, that I audited, was on the social sciences, which, he thought, should be subject to the same methods of testing as the natural sciences. Hempel began his lectures with a famous example from the history of medicine: the story of Ignaz Semmelweis, who during the mid-1840s sought to explain why the incidence of childbed fever was much greater in the first maternity division of the Vienna General Hospital than in the second. This example may be familiar to readers of Hempel since he used it again in 1966, in his classic text *Philosophy of Natural Science*.[1]

Semmelweis invented various different hypotheses and tested them by deriving consequences from each one that could be either verified or refuted by observation. In the first division but not the second, a bell would ring after which a priest would walk through the division to another room to give the last sacrament to a dying

---

1. Carl G. Hempel, *Philosophy of Natural Science* (Englewood Cliffs, NJ: Prentice-Hall, 1966).

woman. One hypothesis Semmelweis considered is that this had a terrifying effect on the women in that division that caused their fever. The hypothesis could be tested by having the priest avoid passing through the division. If the terrifying effect of the priest were the cause, then a consequence of removing this cause should be that the incidence of childbed fever would diminish. It did not, thus refuting the hypothesis. After various such attempts, Semmelweis finally had the idea that the cause might be infectious material being transferred to patients in the first division by physicians who came directly to that division, but not to the second division, after performing autopsies without washing their hands thoroughly afterwards. From the "infectious material" hypothesis, together with the assumption that the physicians in question thereafter will wash their hands very thoroughly with a solution of chlorinated lime before examining patients, it should follow that the incidence of childbed fever will diminish, which is indeed what happened. This observed consequence counted as evidence that the hypothesis in question was true.

Using this example, Hempel proceeds to develop a general view about scientific evidence and method that is called *hypothetico-deductivism*. The basic idea is that the scientist begins with some observed phenomena he seeks to explain. A hypothesis is invented, which, if true, will explain those phenomena. There are no rules of reasoning which the scientist follows in arriving at the hypothesis in the first place; it is simply a guess or conjecture. But once he discovers the idea he should proceed to derive consequences from it that can be directly tested by observation and experiment. The derivation does involve rules of reasoning, viz. deductive ones. As it is sometimes expressed, there is no logic of discovery, only a logic of proof or justification. If the consequences derived from the hypothesis turn out to be false, the hypothesis is refuted as it stands, and must be either modified or rejected. If the consequences turn out to be true, then this fact counts as evidence for the hypothesis. This gives rise to a simple hypothetico-deductive definition of evidence according to which some observed fact e is evidence that a hypothesis h is true if and only if e is derivable deductively from h. There are more sophisticated versions of this idea, as I will show in the chapters that follow. But this will serve my purpose for now.

In his 1954 course, as well as in his 1966 text, Hempel contrasted hypothetico-deductivism with a view that he completely rejected. That view is a version of "inductivism,"

> conceived as a method that leads, by means of mechanically applicable rules, from observed facts to corresponding general principles. In this case, the rules of inductive inference would provide effective canons of scientific discovery; induction would be a mechanical procedure analogous to the familiar routine for the multiplication of integers, which leads, in a finite number of predetermined and mechanically performable steps, to the corresponding product. (p. 14)

According to Hempel, there are no such "mechanically applicable" inductive rules that can be used to generate hypotheses. The idea of a hypothesis may be caused to occur in us by various external and internal factors, but the hypothesis is not inferred from anything, using any rules whatever. It is guessed. Inference enters when it is tested.

When I was an undergraduate I thought I had two alternatives. I could be an inductivist of the sort Hempel describes and advocate the use of rules that don't exist, or I could be a hypothetico-deductivist and encourage the use of imagination to freely invent hypotheses, provided that deduction and observation were employed to test them. Given those options, the choice seemed obvious. Indeed,

PREFACE

Hempel's lectures were enormously exciting and rigorous, so who could resist? He was one of the great teachers of philosophy, as well as being a major twentieth-century innovator in the field.

It was only later, when I began teaching philosophy of science myself, that I discovered other options. Two of my favorite writers, both of whom rejected hypothetico-deductivism, turned out to be scientists, viz. Isaac Newton and James Clerk Maxwell, with very interesting philosophical views about the nature of science. Unlike most scientists, they explicitly discuss and demonstrate in their own scientific work how science should proceed, although their views are quite different from each other. Newton, an inductivist, but, on my interpretation, quite different from the one Hempel describes, always wants evidence sufficient to yield empirical proof of a theory, and presents methodological rules he thinks are needed to get it. Maxwell focuses on situations in which at the moment you can't get evidence sufficient for empirical proof: either you have no theory at all, or you have one but you cannot yet test it experimentally, or you have some evidence but not enough to establish a theory. Without proof, Maxwell insists, there are still important theoretical things you can do. There are scientific methods you can legitimately follow.

In this book I will examine the views advocated by these thinkers, but I will reverse the usual order and begin with evidence, rather than method. In other writings I have distinguished various concepts of evidence used by scientists, and then offered definitions of each. Some of these concepts are objective. In accordance with them, whether some fact is evidence that a hypothesis is true has nothing to do with what anyone knows or believes. There is also a subjective concept of evidence, which is relativized to what a person or group knows or believes. The definitions I offer for these concepts avoid counterexamples to definitions that others have offered. I believe that these concepts, as I define them, are the ones actually

PREFACE

employed in science. The definitions, and the motivation for them, will be developed in chapter 1. (Readers familiar with these definitions from my other works may want to turn directly to section 9 of the chapter, although material in earlier sections does focus more attention than previously on philosophical motivations for my views.) However, an important problem remains. How are these definitions supposed to be applied in individual cases in order to determine whether some fact is evidence that a hypothesis is true? It is one thing to offer a definition, it is another to figure out whether something satisfies that definition in a particular instance. So, it might be said, definition is only half the job.

Enter scientific method. If you want to determine whether some particular fact is evidence for a given hypothesis (in one of the senses of evidence that I define), consult the rules of a scientific method. Newton presents four methodological rules to be used in getting from observed phenomena to a proposition that can be regarded as proved. If these rules are satisfied, will the phenomena count as evidence (in one or more of my senses) that the proposition is true? If so, then perhaps Newton's rules provide a method for determining in particular cases whether some fact e is evidence that some hypothesis h is true, in a sense of evidence that I propose. In chapter 2, I examine Newton's rules and provide a new way of understanding them that, I believe, makes them very plausible and not subject to the many objections that have been raised against them by various scientists and philosophers from the eighteenth century to the present. In chapter 3, I consider several extensions of Newton's rules, a powerful rival methodology, viz. "inference to the best explanation," which I reject, Hume's problem of justifying causal and inductive reasoning of the sort Newton advocates, and finally, the question of whether the concepts of evidence I define in chapter 1 can be applied to particular cases using Newton's rules as I interpret them.

Chapter 4 discusses three methodologies proposed by Maxwell that can be used in situations in which you don't have evidence sufficient to establish a theory, or indeed, in some cases, any theory at all to establish. Philosophers and scientists concerned with general questions of methodology too often focus entirely on methods of proving or establishing theories. But, Maxwell emphasizes, there are more scientific activities involving theorizing than are dreamed of in these philosophies. I think he is right, and that the three different methodologies he presents are important and liberating ones for philosophers as well as scientists to consider.

Hempel's book, to which I referred earlier, was and remains an important text for introducing students to the philosophy of science. Its second chapter, entitled "Scientific Inquiry: Invention and Test," presents in a clear and powerful way two common views of scientific method, hypothetico-deductivism and a type of inductivism. Other authors advocate variations on one of these methods, such as "inference to the best explanation." What I am about to argue is that the methods proposed and used by Newton and Maxwell in their actual work in physics are perfectly reasonable methods, even if neither fits either of the categories Hempel considered or certain variations of them. Indeed, they are superior. Nevertheless, Hempel's work on scientific method, and on a variety of other topics in the philosophy of science, has been inspirational to the present author and to numerous others. It is for this reason that I have dedicated the book to his memory.

Like Hempel, I have tried to make the discussion accessible to a general philosophical reader and not just to specialists in the philosophy of science. There are some more technical sections, but I think one can get the gist of the argument without following the technicalities. In a few sections I have used material from other works I have written.

ACKNOWLEDGMENTS

This book is based on a series of lectures I gave at Seoul National University in the fall of 2011. I want to thank In Rae Cho for inviting me and serving as a wonderful host, and the faculty and students there for philosophical, geographical, and culinary assistance. Talks on individual chapters were delivered at the London School of Economics, University of Pittsburgh, Yeshiva University, Virginia Tech, Ohio University, and University of Münster, and I thank members of the audience for comments and criticisms. The following individuals read the manuscript or portions thereof, and made important suggestions: Kristin Boyce, Linda Brown, Stephen Brush, Victor Di Fate, Steven Gimbel, Marc Lange, John Norton, and Richard Richards. Thomas Wilk gave me expert assistance with the proofs. Finally, I am indebted to the National Endowment for the Humanities for a 2012 Summer Stipend to complete the volume.

EVIDENCE AND METHOD

Chapter 1

A Problem About Evidence

1. Introduction

What does it mean to say that some fact or phenomenon is evidence that a hypothesis is true? Various attempts have been made to define a concept of evidence useful for science. These can be classified in different ways. First, there is an objective-subjective classification. In the objective case the truth of statements of the form "*e* is evidence that *h*" does not depend on who believes or knows what.[1] For example, whether the fact that the planets Mercury and Venus exhibit phases, like our moon, is evidence that these planets revolve about the sun, as Newton claimed in Book 3 of the *Principia*,[2] is independent of whether anyone knows or believes anything. In the subjective case, one natural way to understand this is as elliptical for a statement attributing a belief to Newton; for example, "Newton believed that this fact about the phases of Mercury and Venus is evidence for a claim about their orbits."

---

1. *e* is a sentence describing the putative evidence, *h* is a hypothesis.
2. Isaac Newton, *The Principia*, trans. I. Bernard Cohen and Anne Whitman (Berkeley: University of California Press, 1999), p. 799. Newton, in fact, makes the stronger claim that the heliocentric motion of the planets is "proved by" the fact that they exhibit phases. This kind of proof Newton called a "deduction from phenomena," which he deemed to yield the "highest evidence a proposition can have in this [experimental] philosophy" (letter from Newton to Cotes in 1713; reprinted in H. S. Thayer, *Newton's Philosophy of Nature* (New York: Hafner, 1953), p. 6). In what follows I shall understand this example and others like it from Newton as involving claims about evidence.

EVIDENCE AND METHOD

A second dimension along which theories of evidence can be classified is empirical-a priori. To be sure, claims of the form "*e* is evidence that *h*," where *e* describes some fact, presuppose that *e* is true; and this, in the cases of interest to us, is empirical, not a priori (e.g., that Mercury and Venus exhibit phases). So, more specifically, the empirical-a priori axis should perhaps be thought of as representing claims of the form "*e*, if true, is evidence that *h*." On one view, evidence statements of this form are empirical: they are to be defended (or criticized) by appeal to observationally established facts. On the other view, they are a priori: they are to be defended (or criticized) solely by a priori reasoning such as that involving mathematical or logical computation.

In other writings, I have claimed that although both objective and subjective concepts of evidence are employed in the sciences, the former are the most important.[3] I have also defended the claim that although there are a priori evidence statements, the most important ones are empirical. In this chapter I want to focus on these and raise a fundamental problem. Assuming that there are objective, empirical evidence claims and that we can define "evidence" for them, how are we to apply that definition to particular cases? How are we to determine whether the fact that Mercury and Venus exhibit phases is evidence that they revolve about the sun, in accordance with our preferred definition of evidence? If this were an a priori evidence claim, then a definition of "evidence" might be expected to make reference to the appropriate mathematical or logical apparatus for its application.[4] But if the evidence claim is empirical, how do we

---

3. Peter Achinstein, *The Book of Evidence* (New York: Oxford University Press, 2001); *Evidence, Explanation, and Realism* (New York: Oxford University Press, 2010).
4. This is done, e.g., in Carnap's logical theory of probability and evidence, and in Hempel's satisfaction theory of confirmation. Rudolf Carnap, *Logical Foundations of Probability*, 2nd ed. (Chicago: University of Chicago Press, 1962); Carl G. Hempel, *Aspects of Scientific Explanation* (New York: Free Press, 1965).

determine empirically whether the fact about observed phases of Mercury and Venus is evidence about their revolutions?

Before confronting this question, various philosophical issues about evidence will need to be examined, beginning with the question of whether there are any objective concepts of evidence in use at all, or whether concepts of evidence must all be understood subjectively. After arguing that there are objective concepts, I will reject two assumptions made about them by many writers: the weakness assumption (you don't need much to have evidence), and the a priori assumption (all evidential claims are a priori, not empirical). I will replace these assumptions with three principles I defend that govern evidence and belief, and I will define concepts of evidence that satisfy these principles. These definitions will be contrasted with other objective definitions in the literature, and I will argue that the definitions I propose avoid problems faced by the others. Then, with respect to my definitions, I will pose the fundamental problem raised above: How are the concepts so defined supposed to be applied to determine whether a given fact is evidence for a hypothesis? If there are no general rules for applying the definitions, or if I can supply none, does this mean that the concepts I define are worthless?

2. Is There an Objective Concept of Evidence?

Many writers on evidence certainly think so, whether they take evidence claims to be a priori or empirical. Here is a basic hypothetico-deductive (h-d) view: e (if true) is evidence that h if and only if h entails e. Whether h entails e is an objective fact that does not depend on whether anyone believes or even knows about h or e or their logical relationship. (This view also makes evidential claims a priori, since whether h entails e is decidable a priori.) Fancier versions of the h-d view invoke some idea of explanation. For example, on such views, e is evidence that h if and only if h explains e, or h is the best explanation of e, or h is the most probable explanation of e. On these explanatory versions, whether e is evidence that h is objective, since the concept of explanation invoked is.

Even if such views produce an objective concept, perhaps they are mistaken in thinking that there is such a concept in use, at least in the sciences. Perhaps subjectivists—for example, subjective Bayesians[5]—are right in saying that the subjective concept of evidence is the only one in use in the sciences, or at least that the only reasonable way to understand scientists when they claim, for some e and h, that e is evidence that h, is subjectively.

My initial response to this thought is that those, such as myself, who claim that there is an objective concept of evidence used by scientists need not deny that there is a subjective concept as well—that there is a subjective way to understand certain claims of the form "e is evidence that h." The question is whether there is also an objective way—that is, whether there is an objective concept that reflects what scientists at least sometimes, if not most of the time, have in mind when they speak of evidence, and whether an objective concept is important in understanding and evaluating such evidential claims.

Let's consider a strong subjective view according to which any evidential claim needs to be understood subjectively. On such a view, a statement of the form "e is evidence that h" is to be interpreted as relativized to a person or group: e is evidence that h for person(s) P.

---

5. Bayesians define "evidence" solely in terms of probability. Subjective Bayesians do so using a subjective concept of probability understood as the degree of belief that some person has in a given proposition.

A PROBLEM ABOUT EVIDENCE

And the latter is to be understood in terms of the beliefs of P. Thus, if Newton claims that the fact about the observed phases of Mercury and Venus is evidence that these planets orbit the sun, what Newton means is that this is evidence for him and other astronomers. On a standard Bayesian subjective probability view, this means something like: Newton's rational degree of belief[6] in the proposition that the planets in question revolve about the sun would be increased by his finding out that these planets exhibit phases like our moon. And if we make the same claim as Newton, then we are saying that our rational degree of belief is increased by this fact as well.

An objectivist reply. To be sure, when a scientist makes the claim that e is evidence that h, it can be assumed that he believes that e is evidence that h, or to put it as I did above, that for him e is evidence that h. But this is so for any assertion made by a person. If I assert that p, it doesn't follow that what I have asserted is that, for me, or in my view, p. A subjectivist needs to show otherwise. He needs to show that there is something about evidential claims, as opposed to other objective claims, that makes what is asserted in such claims subjective. He needs to show that such claims can only be understood by reference to beliefs of persons. What kind of argument could there be for such a view?

A subjectivist response. If e is evidence that h, then e provides a good reason for believing h (on my objective view, to be developed later), or (on some views) it provides a better reason to believe h than there is without e. In either case, evidence is related to belief. But how can some e be a good reason to believe h, or a better reason than without e, unless it is a good reason *for someone* to believe h?

---

6. For subjective Bayesians, a set of degrees of belief is rational if and only if it satisfies the axioms of probability. On the subjective view, then, we might speak of a set of degrees of belief as "rationally corrected" to satisfy the probability axioms. It is this set that we attribute to the individual(s) in question.

It can be a good reason for everyone to believe it, or a good reason for some particular person(s) to believe it. Believing is something that humans do, and what is or should be believed is what (some class of) humans do or should believe. This is very different from truth, which has nothing to do with belief. A hypothesis h can be true even if no one believes, or should believe, anything about h.

My response. Medical books provide a list of typical symptoms or signs of a heart attack. If someone exhibits these symptoms or signs, then there is a good reason to believe that person is having a heart attack. That is not sufficient to make the concept of symptom or sign subjective. A symptom or sign of x is not what people believe to be a symptom or sign. It is something that is regularly associated with x, from which x could be predicted, whether anyone knows or believes that it is, or predicts x from the sign. In order to establish whether something is a sign or symptom of x, one need not determine who, if anyone, takes it to be a sign or symptom, or should take it to be; if the concept were subjective we would have to determine this. To be sure, we say that, for certain people, A is a sign of x (e.g., for the typical jury, an accused's refusal to testify is a sign of guilt). But all that means is that some people believe that A is a sign of x. That doesn't make sign a subjective concept. Nor, when I say that something is a sign of x, am I saying that it is a sign *for everyone*—that is, that everyone does or should take it to be a sign. Usually not everyone believes it is a sign; and it is not part of my claim that you or anyone else should take it to be a sign, even when I say it is one, since you or others may not be in an epistemic position to take it to be a sign or even understand what it is a sign of.

The same holds for evidence. If I claim that e is evidence that h, then I am committed to the claim that e is a good reason to believe h (or a better reason than without e, or something like that). But I am not making a claim that for you, or anyone else, e is such a reason.

A PROBLEM ABOUT EVIDENCE

And to determine whether *e* is evidence that *h*, I don't need to establish for which person or persons it is. Nor from the fact that *e* is evidence that *h*, does it follow that *e* is a good reason *for everyone* to believe *h*. Whether it is a good reason for any particular person depends on the epistemic situation of the person. In section 7, I will provide definitions of evidence in terms of objective concepts of explanation and probability that are in accord with such claims.

All of this is not to deny that there is a subjective concept of evidence. When the defense attorney asks the detective in the case about the evidence that the accused committed the crime, he is usually asking about the *detective's* evidence, what he takes to be evidence, what he believes is evidence. This is entirely subjective. His evidence may make him believe that the accused did it, even if it is not reasonable to believe this on the basis of his evidence. This subjective concept of evidence, which is perfectly legitimate, is to be understood as presupposing an objective concept. The detective's (subjective) evidence is what he believes is (objective) evidence. What I am claiming here is: (1) that there is a (legitimate) objective concept of evidence (indeed, more than one), (2) that there is a subjective concept of evidence (perhaps more than one), and (3) that any subjective concept of evidence will need to be understood by reference to an objective one (e.g., that a person's evidence that *h* is to be understood as what that person believes to be evidence that *h*).

Finally, a subjectivist might reply that even if there is an objective concept of evidence, and even if it is to be understood and defined by invoking an objective concept, it is the subjective concept that is usually employed by, and important to, scientists. When Newton claims that the fact that they exhibit phases, like those of our moon, is evidence that Mercury and Venus revolve about the sun, what he really means is that this is what he and other astronomers of the day take to be evidence.

No doubt exhibiting phases is one of the things that made Newton and others believe that Mercury and Venus revolve about the sun. And no doubt in asserting what he did he was implying that he and others believe that this is evidence. But that is not what the claim itself states. The claim itself is not at all subjective or historical but, rather, objective and ahistorical. Indeed, as noted, Newton uses a stronger term than "evidence," viz. "is proved by." Unless the subjectivist can convince me by historical reasons to suppose that Newton meant otherwise, or by philosophical reasons to suppose he should have, I will continue to assume that Newton was making an objective claim about a relationship between the fact that these planets exhibit phases and the proposition in question, and not simply a subjective claim about what he and others believed or should believe.

3. Two Assumptions of Objective Theories

Objective theories of evidence often make two fundamental assumptions that I will reject. The first I call the "weakness assumption":

> *Weakness Assumption*: Evidence is a weak notion. You don't need very much to have evidence for a hypothesis.

By itself this is a vague claim. It will become clearer by looking at some well-known objective theories.

Let's begin with a standard "positive relevance" definition.

> *Positive Relevance*: e is evidence that h if and only if the probability of h, given e, is greater than the probability of h; in symbols, $p(h/e) > p(h)$.[7]

---

7. Carnap, *Logical Foundations*, p. xvi. Carnap (p. 21) calls the positive relevance idea a "classificatory concept" of "evidential support."

A PROBLEM ABOUT EVIDENCE

The basic idea is that e is evidence that h if and only if e boosts h's probability, where the latter is construed objectively. So, on this definition, if

e = I own one ticket in a million-ticket lottery, one winner to be chosen at random
h = I will win,

then e is evidence that h. The fact that I own one ticket is evidence that I will win, since it boosts the probability that I will. To be sure, this is not much evidence, and it is not that strong, but it is to be thought of as some evidence. As the weakness assumption states, you don't need much to have evidence.

Next, a basic "hypothetico-deductive" (h-d) view:

h-d: e is evidence that h if and only if h entails e.

On this view, if

e = light travels in straight lines
h = light consists of particles satisfying Newton's three laws of motion,

then the fact that light travels in straight lines is evidence that Newton's particle theory of light is true, since h entails e, despite the fact that e is also entailed by a contrary theory that light is a wave motion of the sort described by Huygens. In fact on this view, e is evidence for both theories. Again, you don't need very much to have evidence.[8]

Finally, I'll mention a "satisfaction" view of evidence, the most famous version of which is due to Carl G. Hempel in a seminal paper in 1945:

---

8. In chapter 3, I will examine a much more sophisticated version of h-d called "inference to the best explanation."

Satisfaction Theory: e is evidence that *h* if and only if *e* describes instances that satisfy *h* in a sense that can be defined in formal-logical terms.[9]

Hempel defines the "development" of a hypothesis for a class of individuals as what that hypothesis would assert if only those individuals exist. So if our hypothesis *h* is of the form "All *A*s are *B*s," then the development for *h* is a conjunction consisting of sentences of the form

If *x* is an *A*, then *x* is a *B* (for each *x* in the class).

Each individual *x* mentioned in this conjunction "satisfies" the hypothesis "All *A*s are *B*s." Hempel then defines two evidential concepts, "direct confirmation," and "confirmation," as follows:

(a) *e* directly confirms *h* if *e* deductively entails the development of *h* for the class of individuals mentioned in *e*.

(b) *e* confirms *h* if *h* is deductively entailed by the class of sentences each of which is directly confirmed by *e*.

For example, let *e* be that one ball *x*, drawn from an urn containing 1 million balls, is red. Let the hypothesis *h* be that all of the million balls in this urn are red. Then *e* "directly confirms" *h*, since *e* deductively entails the development of *h* for the class of individuals mentioned in *e*. (*e* is "*x* is one of the million balls in this urn, and *x* is red," and the development of *h* for *x* is "if *x* is one of the million balls in the urn, then *x* is red.") So, if Hempel's idea of "confirmation" is associated with that of evidence (as is usually done), then the fact that one ball drawn from an urn containing 1 million balls is red is evidence that all of them are. You need only a sample of one to get evidence for this hypothesis. Evidence, on Hempel's view, is a weak notion.

---

9. Carl G. Hempel, "Studies in the Logic of Confirmation," reprinted in his *Aspects of Scientific Explanation*.

The second assumption made by many (although not all) objective theories of evidence I call the "a priori assumption."

A Priori Assumption: The evidential relationship is a priori, not empirical. Whether *e*, if true, is evidence that *h* (and how strong that evidence is) is a matter to be determined completely by a priori calculation, not empirical investigation.

For example, on one prominent version of the positive relevance view, due to Carnap, whether $p(h/e) > p(h)$, and hence whether *e* is evidence that *h*, is decidable a priori, since whether a probability statement of the form $p(h/e) = r$ is true is, on his view, a matter of a priori calculation. The same holds for the basic hypothetico-deductive view (because of its concept of entailment), for some more sophisticated versions of *h-d* that invoke the concept of explanation (assuming that whether *h* explains *e* is an a priori matter), and for Hempel's satisfaction definition (because of the a priori nature of "development"). According to this assumption, if a scientist presents data described in a statement *e*, and proposes a hypothesis *h*, then whether *e* is true, and whether *h* is true, are empirical issues. But whether *e*, if true, is evidence that *h* is settleable by a priori calculation. A smart philosopher, logician, or mathematician who knew nothing about the science in which *e* and *h* appear could determine whether *e*, if true, is evidence that *h*.

4. I Reject the Weakness Assumption

In all of the examples cited to illustrate the weakness assumption, it seems bizarre, or at least very questionable, to say that *e* is evidence that *h*. In the case of positive relevance, if I own one ticket in a lottery of 1 million tickets, is that really evidence that I will win? Suppose Donald Trump owns the rest of the tickets. Let's put these facts together. Surely this conjunction of facts is evidence that he will win, not that I will, despite the fact that positive relevance is satisfied—that is, despite the fact that both hypotheses receive a probabilistic boost from the combined putative evidence.[10] (I'll call this Example 1.) In the case of the hypothetico-deductive view, even though the Newtonian particle theory entails as well as explains why light travels in straight lines, is that enough to make the rectilinear propagation of light evidence that the Newtonian particle theory is true, especially since the conflicting Huygensian wave theory also entails and explains this? Is it evidence that both theories are true, despite the fact that they are mutually exclusive? (Example 2) Or, finally, referring to Hempel's satisfaction theory, is the fact that one ball selected from the burn is red evidence that all of them are? (Example 3) In Example 1, the probability of the hypothesis is too small for e to count as evidence that h. In Example 3, the sample is too small. And in Example 2, the "evidence" cited doesn't sufficiently discriminate between conflicting theories.

My suggestion is that each of these examples fails to be evidence because in each case the putative evidence fails to provide a good reason to believe the hypothesis in question. And that is what we want evidence for. When Newton cites the observed phases of Mercury and Venus as evidence that these planets revolve around the sun, he is presenting these facts as a good reason for believing the conclusion. The reason that my owning one ticket in a million-ticket lottery is not evidence that I will win is that it does not provide a good reason for believing I will, whereas the fact that Donald Trump owns all but one ticket does provide a good reason for believing he

---

10. If anything, the fact cited is evidence that I will lose.

will win. The reason that the rectilinear propagation of light is not evidence that the Newtonian particle theory that entails it is true is that it provides an equally good reason for believing that the conflicting wave theory is true.

We might put the point by saying that evidence is a "serious" concept. If you tell me that there is evidence that I will win the lottery, then you change my expectations and my behavior: you get my hopes up; I start thinking seriously about how I will spend my fortune; I anticipate. But if your "evidence" turns out to be that I own one of the million tickets, my hopes are dashed; I stop wondering what I will do with the money. To be sure, the fact that I own a ticket boosts the probability that I will win from 0 to 1/1 million. But that's not sufficient to change what I expect will happen or my behavior concerning what will happen. What will change is only the probability of my winning.

A defender of the weakness assumption may reply that in cases of the sort noted above, although we do not have much evidence, we have some, a smidgen, at least more than we had before. Evidence, on this view, is like money. Whether I have $1,000 or $1 in my pocket, I have (some) money in my pocket. If I have any money at all, I have (some) money. Whether I own just one ticket in the lottery, or almost all of them, this is (at least some) evidence that I will win. Evidence doesn't have to provide a good reason to believe a hypothesis, only some reason, however small.

On such a conception of evidence, if the doctor tells you that there is evidence, at least a tiny bit, that you have cancer, you will probably get quite worried, understandably so. But if the putative evidence is simply that you have a stomach ache (which boosts the probability of this hypothesis a bit), and that is all the "evidence" the doctor has, then I think you have no cause for a cancer concern; you would be, and be considered, a worrywart at best, a hypochondriac

EVIDENCE AND METHOD

at worst, if you had such a concern. Indeed, my advice is to fire your doctor. The stomach ache by itself is no reason at all to believe that you have cancer, not even a little reason.

Some concepts are "threshold" concepts. There needs to be a certain amount of something for the concept to apply. For example, "hirsute" and "wealthy" are threshold concepts with respect to hair and money. Not just any amount of hair will make you hirsute, not just any amount of money will make you wealthy. If you have just five hairs on your head, you are not hirsute, not even a little bit; indeed, you are bald. If you have just $5 to your name, you are not wealthy, even a little bit; indeed, you are poor. What the threshold amount needs to be can be rather vague and imprecise; or for legal or other purposes it can be made precise (for tax purposes the government might set a threshold number for being wealthy, as it does for poverty). I am suggesting that "evidence" is a threshold concept, rather than a "continuous" one. There needs to be a certain amount to count as evidence, even a little bit of evidence.

A certain amount of what? Later, in developing my own definitions of evidence, I will take this to be an amount of probability. I will say that evidence is a threshold concept with respect to probability (just as being wealthy is a threshold concept with respect to money). Fact e is evidence that h is true only if the probability of h, given e, exceeds the threshold (whatever this turns out to be); otherwise e is not evidence that h, even if the probability of h, given e, is greater than zero, and even if it is greater than the probability of h without e. The reason that it seems so questionable to count the lottery and other examples above as providing evidence for the hypotheses in question is that this probability threshold is not reached. A probability of 1/1 million is not enough to count as any evidence at all that I will win the lottery. When I buy a ticket to the lottery, there is a boost in something. But that something is probability, not evidence. As I said earlier, evidence is a "serious" concept. In terms of probability, this means that the probability of *h* on the putative evidence *e* must be sufficiently high to make it reasonable to believe *h*, given *e*. It must be sufficiently high to yield any evidence at all.

5. I Reject the A Priori Assumption

My view is that, in general, when an objective evidence claim is made, it is empirical. By this I mean that it is, and ought to be, defended (or criticized) by empirical arguments, not (just) a priori ones. To be sure, some evidence claims might be construed as a priori, if sufficient assumptions are packed into them. For example, the following claim might be taken to be a priori: the fact that I own 95 percent of the tickets in this fair lottery, one ticket of which will be drawn at random, is evidence that I will win. (Later, however, I will question even this example.) Contrast that with the following claim: the fact that I own 95 tickets in this lottery is evidence that I will win. Whether it is or isn't depends on empirical facts about the lottery, such as whether it is fair, and what percentage of the tickets I own. If I claim that the fact that I own 95 tickets in the lottery is evidence that I will win, and I am asked to defend this claim, I would appeal to empirical facts, such as that the lottery is fair and that my 95 tickets represent 95 percent of all tickets. I wouldn't just do some calculations.

In Galileo's *Dialogue Concerning the Two Chief World Systems* (1632),[11] there is an Aristotelian evidential claim considered and defended by the character Simplicio. The hypothesis being considered

---

11. Stillman Drake, trans., 2nd ed. (Berkeley: University of California Press, 1967).

is that the earth is stationary and does not turn on its axis. The evidence for this is that when heavy objects fall they fall straight down by a vertical path to the earth. Here is a relevant passage (p. 126).

> As the strongest reason of all [for the stationary earth hypothesis] is adduced that of heavy bodies, which, falling down from on high, go by a straight and vertical line to the surface of the earth. This is considered an irrefutable argument for the earth being motionless. For if it made the diurnal rotation, a tower from whose top a rock was let fall, being carried by the whirling of the earth, would travel many hundreds of yards to the east in the time the rock would consume in its fall, and the rock ought to strike the earth that distance away from the base of the tower [which, of course, doesn't happen].

For our purposes, the evidential claim is this: the fact that a heavy falling body (such as a rock falling from a tower) falls vertically (and reaches the base of the tower) is "irrefutable" evidence that the earth is motionless. Note that the Aristotelian doesn't just make this evidential claim but goes on to defend it as well, and does so empirically by saying what would happen if the earth were rotating (the rock would land a distance away from the base), and (again empirically) that this does not happen. To be sure, after presenting the Aristotelian's defense of the evidential claim, Galileo (using the character Salviati) proceeds to argue that the evidential claim is false, since it is based on a false empirical assumption about what would happen if the earth were rotating (the false assumption is that the rock would fall a distance away). But my point is the same. The claim that the fact that the rock falls to the base of the tower is evidence that the earth is motionless is being defended by the Aristotelian, and attacked by Galileo, by reference to empirical assumptions

A PROBLEM ABOUT EVIDENCE

about motion. It is not defended, or attacked, by a priori means. The evidential claim itself, then, is being treated as an empirical claim, not an a priori one.[12]

The same is true of earlier examples. Suppose I claim that the fact that I own 95 tickets in the lottery is evidence that I will win. I might defend this claim by citing the empirical fact that there are 100 tickets in the lottery, one of which will be chosen at random. You might attack my evidential claim by trying to show empirically that my defense is misguided since there are, in fact, 100,000 tickets in the lottery, not 100. Or, to take a different example, if I claim that the fact that the ball I drew from the urn is red is evidence that all of them are, I might defend this empirically by claiming that all the balls in the urn have the same color; you might attack my evidential claim, again empirically, by arguing that the balls in the urn are not uniform in color.

An a priorist will have a reply. Let us call an evidence statement of the form "*e* is evidence that *h*" *empirically complete* if there is sufficient information in *e* to make the evidence statement true a priori. The a priorist will then say that any empirically incomplete evidence statement should be revised in such a way as to make it empirically complete. What scientists should aim to do, or at least what philosophers should help them do, is to make their evidence statements empirically complete. If this is accomplished, then not only will we

---

12. In fact, the claim that the rock falls to the base of the tower is an approximation or idealization, since, if the earth moves, then the horizontal motion of the top of the tower is greater than that of the base, since both the top and the base have to turn with the earth once in 24 hours; so the top has a greater distance to go than the base during that period. Hence, the horizontal motion of the rock will be greater than that of the base, causing the rock to land slightly to the east of the base. Galileo realized this, but given the tiny size of the tower compared with the radius of the earth, he was unable to measure this small displacement. It was empirically demonstrated by Johann Benzenberg in 1802, in a tower in Hamburg, and in 1804, in a mine shaft in Schlebusch. I am indebted to Stephen Brush for the point and the reference.

have evidential claims that are true but ones whose truth can be demonstrated by a priori calculation. So, for example, in the lottery case, my evidence should not be taken to be just that I own 95 tickets, but that there are 100 tickets in all, one of which will be selected at random. This will turn my original empirically incomplete evidence statement into a complete a priori claim.

Will it? Well, suppose it is a rule of the lottery that no one can win who is an official of the lottery; and suppose I am an official of the lottery. If, as seems plausible, these empirical facts would suffice to refute my revised (as well as my original) evidential claim, then that claim is not empirically complete. Indeed, if that claim can be refuted by empirical means, then it is not a priori at all. Let's try to make my lottery claim empirically complete. Suppose, then, as it turns out, I am not an official of the lottery, and I add this fact and the fact that no official is eligible to win to my revised evidence claim. Is this new claim empirically complete? Is it true a priori? What if the lottery is illegal in the state, so that no one is permitted to win? We can keep this up indefinitely (suppose that for some unforeseen circumstance a drawing will never be made, etc.).

An a priorist may reply that an evidential claim is always a priori, whether or not it is empirically complete, and whether or not it is a priori true or false. My original evidential claim is a priori false, and if I want to revise it in such a way as to make it a priori true, I must make it empirically complete. If this is the response, then what is happening is that an appeal is made to an empirical fact (about my status as a lottery official) that must be added in order to (try to) change my original evidential claim from an a priori false one to an a priori true one. A defender of an empirical concept of evidence will say that whether we understand my original evidential claim as a priori false or empirically false, an appeal will need to be made to empirical facts in order to turn it into a true evidential claim.

A PROBLEM ABOUT EVIDENCE

For the sake of argument, assume that we could turn the original evidential claim into an empirically complete one. What is the advantage in doing so? I can hear the a priorist say things like: decisiveness, certainty, general agreement. As with mathematical calculations, there is a right answer that all can agree on, there is proof. So we can settle disputes that frequently arise in the sciences and elsewhere about whether some putative evidence that h really is evidence that h. This is one of the advantages of an a priori view cited by Carnap.

Even if the evidential claim is made a priori, that does not guarantee decisiveness, certainty, or general agreement. Philosophers notoriously disagree about whether various statements are a priori true (e.g., "every event has a cause"), or even whether they are a priori. Why should we expect anything different with empirically complete evidence statements? But even if, as with Carnap, evidence statements are interpreted as statements about probability (e.g., as $p(h/e) > p(h)$), and the latter are construed as a priori, the calculations needed to determine such probabilities may be very difficult if not impossible to make. (Even with Carnap's simple languages, the computations can become staggering.) Determining the truth of an empirical evidence claim may be much simpler, and lead to more agreement, than determining the truth of some empirically complete (a priori) revised version of it.

Most important, even if we could arrive at empirically complete evidence statements, consider what we have to do to get them. We need to include in the revised evidence statements themselves empirical facts that might otherwise be appealed to in defense of the "original" evidence claims, should we choose to defend them. And we need to include ones sufficient to make the resulting claims a priori and true. Indeed, as the above lottery example shows, no matter how many revisions we make, a clever objector will probably come up with a new empirical condition that shows that the ones cited are not empirically complete. Can a scientist or a philosopher think of all possible defenses of the original evidence claim so as to make the revised one complete? I doubt it.

In any case, this a priori ("Cartesian") undertaking is not how science in fact operates, or needs to. Whether an evidence claim, construed empirically, needs to be defended, and how, is a contextual matter that depends on the kinds of questions likely or important in the context. Both the a priorist and the empiricist (as I shall call a defender of an empirical concept of evidence) agree that evidence claims can and should be defended. They both agree that empirical information needs to be invoked (the one advocates putting it inside the evidence statement, the other invoking it in its defense). The difference is that the a priorist, but not the empiricist, requires that evidential claims be a priori, and therefore that any empirical facts cited in defense of an evidential claim be sufficient to transform that claim into an a priori one. According to the empiricist, an evidential claim can be defended empirically without making the relationship between the defense and evidential claim a priori. In a given context it may be perfectly appropriate for me to defend my claim that the fact that (e) I own 95 tickets in the lottery is evidence that (h) I will win by saying that (d) there are 100 tickets in the lottery. This is so, even though defense (d) by itself does not entail that e is evidence that h, as the counterexamples above show.

Finally, when scientists themselves defend or attack an evidential claim, the defense or attack is usually empirical, not a priori. As noted above, the Aristotelian defends the claim that the path of the falling rock is evidence that the earth is stationary by appeal to the (alleged) empirical fact that if the earth were turning and the rock were falling from the top of the tower, the rock would fall behind the turning tower, which it doesn't. Galileo attacks the evidential claim by appeal to the empirical claim that if the earth were turning and the rock falling, its motion would be a compound of two motions, not one. Neither appeal by itself entails the truth (or falsity) of the original evidential claim. Other assumptions would need to be invoked. But why do so if the defense or attack is sufficient for the context of inquiry? The Aristotelian's evidential claim is refuted not by showing that he has made some a priori mistake of calculation, but by showing, or arguing, that he has assumed true an empirical claim (that the falling rock has only one motion relative to the earth) that he has no right to assume (unless he already assumes that the earth is stationary). The a priorist must say that a defense is never complete until it shows an a priori relationship between the defense and the proposition defended. An empiricist makes no such demand.

6. Three Principles of Objective Evidence

If we reject both the weakness and a priori assumptions for objective evidence, what principles should govern such a concept? I will propose three principles for objective evidence in the present section, followed by some definitions of evidence in the next section. The first principle I call simply

> *The First Principle of Reasonable Belief:* If e is evidence that h, then e is a good reason to believe h.

All of the counterexamples cited earlier against the positive relevance, hypothetico-deductive, and Hempelian satisfaction definitions violate this principle. That is why they are counterexamples. They provide conditions that are too weak to count something as evidence. They do so, I have claimed, because they allow us to classify something as evidence that h is true even though it does not provide a good reason for believing h. Indeed, following my idea that evidence is a "threshold" concept with respect to probability, they allow something to be evidence that h is true even though it provides no reason at all to believe h. It may increase the probability of h (positive relevance), it may be entailed by h (hypothetico-deductive), and it may "satisfy" h (Hempel), but it can do these things without reaching a threshold of probability necessary to make h reasonable to believe.

How can we decide what that threshold is? To help us get to that decision, I propose a second principle:

> *The Second Principle of Reasonable Belief:* If e is a good reason to believe h, then it cannot be a good reason to believe a hypothesis incompatible with h.

If the fact that light exhibits interference patterns is a good reason to believe that light is a wave motion (as wave theorists Young and Fresnel thought in the nineteenth century), then it cannot also be a good reason to believe that light consists of Newtonian particles, even if both theories can explain interference, even if interference is "satisfied" by both theories (in something analogous to Hempel's sense), and even if interference gives some probabilistic boost to both theories. The fact that a coin has been tossed increases the probability from 0 to .5 that it will land heads. But if that is a good reason for believing it will land heads, then it is an equally good reason for believing it will land tails, since the probability of that hypothesis has also increased from 0 to .5. It's an equally good reason for believing both hypotheses, and hence not a good reason for believing either one. To be sure, if a coin has been tossed, that is a good reason to believe it will land heads or tails, but it is not a good reason to believe either one. And if I buy one ticket in a million-ticket lottery, that increases the probability that I will win, and it is a good reason to believe I am now eligible to win, but it is not a good reason to believe I will win.

The third principle governing objective evidence derives from my rejection of the a priori assumption in the previous section:

> *Empirical Principle of (Objective) Evidence:* In general, whether e, if true, is evidence that h is an empirical, not an a priori, question.

I say "in general" because I do not want to preclude the possibility of formulating empirically complete, and hence a priori true, evidential claims. But, as I said in the previous section, I see no particular need to do so, and, in general, evidential claims actually made in science satisfy this principle. They are subject to an empirical defense or criticism.

7. Definitions of Evidence

Using these principles as guidelines, let me proceed to define several concepts of evidence satisfying them. According to the first principle, evidence e must be a good reason to believe h. My suggestion is that e is such a reason only if the probability of h, given e, is sufficiently high. This corresponds to one of Carnap's concepts of evidence, viz., that e is evidence that h if and only if $p(h/e) > k$, where k is some threshold value for "sufficiently high." Carnap does not supply any particular number or range for k; and he treats this condition as both necessary and sufficient for evidence (in this sense).[13]

---

13. As noted, Carnap also recognizes a second concept of evidence, positive relevance, according to which e is evidence that h if and only if $p(h/e) > p(h)$.

By contrast, I am proposing this only as a necessary, but not a sufficient, condition. It remains to say what else is necessary, what number or range k represents, and what concept of probability is being employed.

First, the question of k. According to the second principle of reasonable belief, if e is a good reason to believe h, then e cannot be a good reason to believe a hypothesis incompatible with h. From this and the first principle of reasonable belief, it follows that if e is evidence that h, then e cannot be evidence that h' where h' is incompatible with h. Given this and the probabilistic assumption above, it follows that k must be greater than or equal to one-half. If k were less than one-half, then, since incompatible hypotheses can both have probabilities less than one-half, it would be possible for some e to be evidence for each of two incompatible hypotheses, which violates the second principle. Therefore, we may conclude that e is evidence that h only if $p(h/e) > \frac{1}{2}$. Another way to put this is that e is evidence that h only if h is more probable on e than its denial is on e: $p(h/e) > p(-h/e)$.

What else is necessary? Suppose that the probability of h, given e, is greater than one-half. This can be so even if, as we might say, e has nothing to do with h. For example, let

e = John F. Kennedy was the 35th U.S. president
h = the law of inertia

The probability of h is very high, given e (since it is high without e, and e doesn't affect h's probability). Yet e is not evidence that h is true. Some may want to say that the reason e is not evidence that h is that e does not boost h's probability, which is necessary for evidence. But that won't help, because boosting the probability of a hypothesis is not necessary for evidence.

A PROBLEM ABOUT EVIDENCE

To show this, consider another example. Suppose I have some unpleasant symptoms S, and I have taken a drug D that works to relieve those symptoms 95 percent of the time. Call this background information b, and let h be the hypothesis that my symptoms will be relieved. Then, given this background information b, the probability of the hypothesis h is very high—say, .95. That is, $p(h/b) = .95$. Now suppose that five minutes later I decide to take another drug D' which also relieves symptoms S. Drug D' is slightly less effective than D, but still quite effective—say, 90 percent. Its advantage is that it has fewer side effects than D, and when taken within 15 minutes of taking D it completely cancels the effective power of D without destroying any of its own. Let e be this information about taking D'. The probability of h, given e and b, is, let us say, .90. Even though $p(h/e\&b) < p(h/b)$, surely e (the fact that I have taken drug D', which, while it cancels the effect of drug D that I took, is still 90 percent effective in reducing symptoms S) is evidence that my symptoms will be relieved. This is so, even though the probability of my experiencing relief has decreased. Indeed, the fact that I have taken D' is a very strong reason for believing that h is true.[14]

Why is Kennedy's being the 35th president not evidence that the law of inertia is true (despite the fact that the probability of h, given e, is very high), whereas my taking drug D' is evidence that my symptoms will be relieved (despite the fact that the probability of this hypothesis is thereby decreased)? In both cases, the probability of h, given e, is very high. What is the difference? The fact about John F. Kennedy, we might say, has "nothing to do with" the law of inertia, whereas the fact that I am now taking drug D' does have "something

---

14. Not only is boosting the probability of a hypothesis not necessary for evidence, it is not sufficient either, as was argued in section 4.

to do with" the relief of my symptoms. My suggestion is that this "something to do with" is a matter of an explanatory connection, or rather the probability of such a connection, between the hypothesis and the evidence. Let me explain.

I will say that there is an explanatory connection between h and e if and only if either (a) h correctly explains why e is true, or (b) e correctly explains why h is true, or (c) some hypothesis correctly explains why both h and e are true. An example of (a) is a case in which e is the fact that this body is accelerating, and h is the hypothesis that a force is being exerted on the body. An example of (b) is a case in which e is the fact that I own 95 percent of the tickets in the lottery and h is the hypothesis that I will win. An example of (c) is a case in which e is the fact that, in the first 100 tosses, this coin landed on heads all of the time, h is that it will land on heads next time (where both h and e are correctly explained by the hypothesis that the coin has a very strong physical bias toward heads).[15] On my proposal, then, e is evidence that h only if, given e, it is probable that there is an explanatory connection between e and h. How probable? Well, at least more probable than that there is no explanatory connection. So, writing $E(h,e)$ for "there is an explanatory connection between h and e," we have

---

15. In the last case, suppose we change the example by adding the information b that the coin is not physically biased, is not two-headed, and that the previous 100 tosses were made in perfectly randomized way without any foul play. In such a case, given b, e (that the coin landed heads 100 times) is not evidence that h (it will land heads next time). In virtue of b, e is not a good reason for believing that h will be true. This is because in virtue of b, we are dealing with outcomes of a coin toss that are probabilistically independent. They are so because there is no explanatory connection between h and e (or rather the probability of such a connection is very low). It is just a fluke that we got 100 heads in a row, but it can happen. After all, the probability of its happening is not zero, but $½^{100}$. When, given e, the probability of an explanatory connection between h and e is low or zero, e is not evidence that h. I thank Wendy Parker for pressing me on this issue.

A PROBLEM ABOUT EVIDENCE

$$e \text{ is evidence that } h \text{ only if } p(E(h,e)/e) > 1/2. \qquad (1)$$

That is, e is evidence that h only if the probability that there is an explanatory connection between h and e, given e, is greater than one-half.

Now "there is an explanatory connection between h and e" entails that h is true. (h cannot correctly explain why e is true unless h is true; the same holds for the two other possibilities for an explanatory connection given in the definition.) It follows from (1) and the rules of probability that

$$e \text{ is evidence that } h \text{ only if } p(h/e) > 1/2, \qquad (2)$$

which is the "threshold" idea proposed earlier.[16] Although (1) entails (2), the reverse is not the case, as the Kennedy example shows. The probability is high that the law of inertia is true, (even) given that Kennedy was the 35th president. But given that he was the 35th president, the probability is not high that there is an explanatory connection between this fact and the law of inertia. (It is not probable that the law of inertia correctly explains why Kennedy was the 35th president, or that his being the 35th president correctly explains why the law of inertia holds, or that something correctly explains both why the law holds and why he was the 35th president.) In this case, e and h probably have "nothing to do with each other" in the sense I have suggested. That is why e is not evidence that h, despite the fact that the probability of h, given e, is high—that is, despite the fact that (2) is satisfied. (1) is not.

By contrast, in my drug case, (1) in addition to (2) is satisfied. Given the fact that I have taken drug D', there is probably an explanatory

---

16. (1) can be shown to be equivalent to: e is evidence that h only if $p(E(h,e)/e\&h) \times p(h/e) > \frac{1}{2}$. The latter entails (2). See *The Book of Evidence*, p. 153.

connection between my symptoms being relieved and my taking drug D'. In this case, it is probable that the reason my symptoms will be relieved is that I have taken drug D'.

David Hume famously claimed that when we make an inductive inference from an observed set of facts to a generalization, or to another fact not yet observed, we suppose that there is some causal connection between the two. (Hume argued that we are never justified in doing so, and hence that inductive inferences are never justified. I will leave a discussion of Hume's famous problem until chapter 3.) I put Hume's causal claim in terms of an explanatory connection (which can include causal-explanatory ones). I apply it to cases of evidence (which includes cases involving inductive generalization, but others as well). And I add a dollop of probability. (Some state of affairs is evidence that some other state of affairs obtains only if, given the former, there is probably an explanatory connection between the two states of affairs.)

To complete my definition of evidence, I will add two simple conditions. The first is that e is true. The fact that this body is accelerating cannot be evidence that there is a force being exerted on it if it is not accelerating. We could say, counterfactually, if this body were accelerating, that would be evidence that a force was being exerted. More generally, we could say that e, if true, is evidence that h. But the claim that e is (in fact) evidence that h presupposes the truth of e.

Second, I suggest that if e entails h (logically or semantically), then e is too "close" to h to be evidence that h. The fact that the president is wearing a blue tie is not evidence that he is wearing a tie. The fact that force = mass × acceleration is not evidence that force is proportional to acceleration. The fact that John is married is not evidence that John is not single. Some philosophers want to say that entailment is "ideal" evidence, or that it is a "limiting case" of evidence. Those who speak this way seem to be suggesting that this is

A PROBLEM ABOUT EVIDENCE

the best or strongest sort of evidence possible.[17] By contrast, I regard it as an evidential analogue of a circular argument in logic—the idea that you are assuming what you are trying to show, or something close to what you are trying to show. Although "today is Tuesday" does entail "Tuesday is a day," an argument with the same premise as the conclusion, or with one too close to the conclusion, is unconvincing, even if valid. The same holds true, I suggest, with evidence.[18]

Putting these conditions together we have:

e is evidence that h only if
(i) $p(\mathrm{E}(h,e)/e) > \frac{1}{2}$,
(ii) e is true,
(iii) e does not entail h.

If these conditions are satisfied, I say that e is "potential" evidence that h. If in addition

(iv) h is true,

I speak of this as "veridical" evidence. If, in addition, it is the case that

(v) $\mathrm{E}(h,e)$,

---

17. Even if we were to allow e to entail h as a limiting case, it would not follow that e is evidence that h. The first (explanatory) condition needs to be satisfied. And from the fact that e entails h it does not follow that, given e, there is probably an explanatory connection between e and h.

18. To be sure, this condition allows e plus something else to entail h. For example, the fact that this body is accelerating (e) is evidence that a force is being exerted on it (h). e plus Newton's second law entails h, but e by itself does not. The present no-entailment condition does not allow this conjunction to count as evidence that h, which seems right to me. In defense of the claim that e is evidence that h, we could appeal to Newton's second law. Equivalently, we can say that, in view of the fact that Newton's second law holds, e is evidence that h. But that is different. When we conjoin e with Newton's second law we get a proof of h, or an argument demonstrating that h is true. We get something stronger than evidence. When we defend the claim that e is evidence that h by appeal to Newton's second law, we get a defense of an evidential claim, not an expanded evidential claim.

that is, that there is an explanatory connection between h and e, I will say that e is "strong veridical" evidence that h.

For example, let e be that I own 95 percent of the tickets in this lottery, let h be that I will win. Suppose that both h and e are true, and that there is an explanatory connection between h and e (e.g., the reason I will win is that I have 95 percent of the tickets). Then e is strong veridical evidence that h. Suppose that condition (v) does not hold (e.g., I will win not because I own 95 percent of the tickets, but because, even though very improbable, there was in fact cheating going on in my favor). Then e is veridical evidence (since h is true), but not strong veridical evidence. Suppose h turns out to be false (even though there is no cheating, one of the 5 percent of the tickets I don't own will win). Then e is potential but not veridical evidence that h.

On my view, when scientists assert evidential statements they intend to be making strong veridical evidential claims—that is, ones satisfying all of the previous conditions. When Simplicio asserts that the fact that (e) the rock falls to the base of the tower, rather than at some distance from it, is evidence that (h) the earth is motionless, he intends to be making a strong veridical evidential claim. As Salviati argues, he fails in this attempt since, given e, the probability of an explanatory connection between h and e is not high (because of Galilean relativity). So e is not even potential evidence that h. Nor, of course, for the same reason, will e be potential evidence for Galileo's hypothesis that the earth moves. Galileo needs to provide other information that will count as evidence for this hypothesis. My claim is only that what scientists aim for when they seek evidence is information e that will provide a good reason to believe a hypothesis h—in a sense of "good reason" that requires not just the high probability of an explanatory connection between h and e but also that there really is such a connection, and hence that both h and e are true.

A PROBLEM ABOUT EVIDENCE

Frequently, scientists who offer evidence fail to provide either veridical or potential evidence. When they do so, we can still speak of what they present as evidence—for example, as Simplicio's (or Aristotle's) evidence that the earth is motionless. One way to understand the latter is in a purely subjective sense, as what Simplicio (or Aristotle) believed (viz. that e is strong veridical evidence that h). More generally, I will say that

e is X's *subjective* evidence that h if and only if X believes that e is (strong) veridical evidence that h, and X's reason for believing h is true is that e is true.

Another way to relativize an evidential claim is to an epistemic situation,[19] saying that anyone in such a situation would be justified in believing the hypothesis in question. I will call this ES (epistemic situation)-evidence, and say that

e is ES-evidence that h (with respect to an epistemic situation ES) if and only if e is true and anyone in ES is justified in believing that e is (strong) veridical evidence that h.[20]

All of the concepts of evidence I have introduced are different from what are called Bayesian concepts of evidence, as well as from a concept that might be associated with "inference to the best explanation." Bayesian concepts are defined entirely in terms of probability,

---

19. This is a type of abstract situation in which one knows or believes that certain propositions are true, and one is not in a position to know or believe that others are, even if such a situation does not in fact obtain for any person. For more on this idea, see *The Book of Evidence*, pp. 20–21.
20. e could be X's subjective evidence that h without being ES-evidence that h relative to X's epistemic situation, since no one in ES may be justified in believing that e is veridical evidence that h. For a more extensive discussion of these concepts of evidence, see *The Book of Evidence*, chaps. 1 and 8.

and introduce no ideas about explanation (for example, e is evidence that h if and only if $p(h/e) > p(h)$). "Inference to the best explanation" does introduce the concept of explanation, but at least in the version due to Peter Lipton, its most dedicated recent proponent, it does not introduce probability.[21] According to Lipton (whose views will be critically discussed in chapter 3), when we make an inference from phenomena described in e to some hypothesis h, we should understand this as an inference to a hypothesis that furnishes what he calls the "loveliest" explanation of e (one that, if correct, will "provide the most understanding," rather than to a hypothesis that gives the "likeliest" (most probable) explanation of e). If we employ Lipton's ideas to form a definition of (potential) evidence, we would have: e is evidence that h if and only if h, if true, is the "loveliest" explanation of e. Unlike my concept of potential evidence, this does not introduce the idea that the explanation must be probable.[22] In any case, my definition requires that it be probable that there is an explanatory connection between h and e, not that it be probable that h explains e (it could be probable that e explains h or that something explains both h and e).

8. Do the Definitions Yield Objective Concepts of Evidence Satisfying the Three Epistemic Principles?

To begin with, let me say something briefly about the concepts of probability and explanation that I am using in my definitions of evidence. The former concept is an objective probabilistic measure of

---

21. Peter Lipton, *Inference to the Best Explanation*, 2nd ed. (London: Routledge, 2004).
22. Lipton does claim that if h gives the "loveliest" explanation of e, then, in general, it will also give the "likeliest." But he does not show why this must be so. And, as I will indicate in chapter 3, no argument that I can discover shows that this is so.

A PROBLEM ABOUT EVIDENCE

the degree of reasonableness of believing something. Certain physical (and/or mathematical) facts or states of affairs may make it reasonable to a certain degree to believe something. Suppose that this coin, which has two sides marked heads and tails, is balanced and will be tossed randomly three times. There are eight possible outcomes, three of which involve exactly two heads. These facts make it reasonable to the degree 3/8 to believe that the coin will land heads exactly twice. That there is this degree of reasonableness is an objective, nonphysical, normative fact determined by the former physical and mathematical facts. It does not depend on what anyone knows or believes. Assuming that the initial suppositions are correct, if you, or the experts, believe that the degree of reasonableness (i.e., the probability) is ½, you and the experts are just wrong. The fact that an unbalanced force is acting on some body makes the degree of reasonableness of believing that the body is accelerating very high—for example, much greater than ½—whether or not anyone believes this, or knows anything about accelerations or Newton's second law.

Probability construed objectively as degree of reasonableness of belief is what I am using in my definitions of evidence. I call it "objective epistemic probability."[23] This concept may be employed in making a priori or empirical probabilistic claims. If the coin being considered in the first example is not a real one but an abstract "ideal" one tossed in an "ideal" random way, then the associated probability claim is a priori. If a real coin is being referred to, then the associated probability claim is empirical, as it is in the case of the second example involving acceleration.

The concept of explanation I am employing is also objective. In other works I define a concept of "correct explanation," which is noncontextual and does not depend on the beliefs or purposes of

---

23. See *The Book of Evidence*, chap. 5, for an extensive discussion of this concept.

anyone giving or receiving the explanation.[24] (This I distinguish from a concept of "good explanation," which does so depend, and may or may not be a correct explanation.) With the concept of a "correct explanation," I argue, whether something satisfies the concept is an empirical, not an a priori, question.[25]

With these concepts of probability and explanation, we can conclude that it will be an objective and (usually) empirical matter whether e is potential or veridical evidence that h. Whether e is X's (subjective) evidence that h depends on whether X believes that e is veridical evidence that h, which is an empirical but subjective fact about X. Whether e is ES-evidence that h depends on whether anyone in ES is justified in believing that e is veridical evidence that h, which, depending on how ES is described, can be objective and empirical.[26]

Do these concepts of evidence satisfy the three principles at the beginning of section 6? The first principle is that if e is evidence that h, then e is a good reason to believe h. If the probability of h, given e (i.e., the degree of reasonableness of believing h, given e) is high, then there is a good reason to believe h. For e to be evidence that h, more is required than simply that h is probable given e; it is required that e have "something to do with h." My ("Humean") idea is that this "something to do with" be understood in terms of an explanatory

---

24. *The Book of Evidence*, pp. 160–66; *The Nature of Explanation* (New York: Oxford University Press, 1983), pp. 103–6.
25. Nancy Cartwright follows me in requiring an explanatory connection between evidence and hypothesis. By contrast to me, however, she takes explanation as primitive, and does not believe that a definition is needed. See her essay "Evidence, External Validity, and Explanatory Relevance," in Gregory J. Morgan, ed., *Philosophy of Science Matters: The Philosophy of Peter Achinstein* (New York: Oxford University Press, 2011).
26. To be sure, if an ES is described as some particular person's ES, then whether that person is in that epistemic situation is a subjective, and empirical, fact about him. But given that he is in that ES, whether anyone in that epistemic situation would be justified in believing that e is veridical evidence that h is an objective fact. If we were to describe such an ES so as to make it "empirically complete" with respect to h, the issue would become a priori. But even with the best efforts of historians of science this task would be very difficult.

A PROBLEM ABOUT EVIDENCE

connection between *e* and *h* (or the probability of such a connection). How high should the probability be? The second principle is that if *e* is a good reason to believe *h*, then it cannot be a good reason to believe a hypothesis incompatible with *h*. So from the first and second principles, if *e* is evidence that *h*, then it cannot be evidence that *h'* if the latter is incompatible with *h*. If we make it a requirement for evidence that the probability of an explanatory connection between *h* and *e*, given *e*, be greater than ½, then the first and second principles are satisfied. On my definitions, both potential and veridical evidence satisfy this requirement and therefore these principles. ES and subjective evidence do not. Someone may believe that *e* is veridical evidence that *h*, and, given his epistemic situation, may even be justified in that belief, without its being the case that *e* is a good reason to believe *h*.

Should the probability requirement for potential and veridical evidence call for an even higher probability? One possibility is to say that the probability threshold should be significantly greater than ½, but that there is no precise threshold value that is both necessary and sufficient for evidence. A threshold value of ½ is necessary but not sufficient. A threshold value of .95 is sufficient but not necessary. No number is both. If I own 5 percent of the lottery tickets, this is not evidence that I will win; if I own 95 percent, it is evidence. What if I own 55 percent? There is a gray area with no precise boundaries in which there is no definitive answer. On this view, evidence is a threshold concept with respect to probability in the following sense: although a probability > ½ is a necessary condition, there is no probability value that is such that all probabilities higher than it satisfy the probability requirement and all lower probabilities fail to satisfy it. There is some essential vagueness.[27]

---

27. An analogy might be with the concept of "intelligent" with respect to IQ. One idea is to set an IQ of 100 as a necessary threshold value for being intelligent, since if your IQ is better than that you are more intelligent than approximately half the population, but to have no threshold that is both necessary and sufficient.

A second possibility is to leave ½ as a threshold value for evidence that is both necessary and sufficient. This generates a more precise concept of evidence, albeit one that is weaker. For the sake of simplicity I will retain ½ as a necessary and sufficient threshold value.[28]

The third principle at the beginning of section 6 is that, in general, whether e, if true, is evidence that h is an empirical, not an a priori, question. As already noted, this is satisfied by all of my definitions of evidence, since, in general, whether the probability is high that there is an explanatory connection between e and h, given e, is an empirical question for veridical and potential evidence. Whether e is X's subjective evidence depends on the empirical fact about what X believes. And whether e is ES-evidence that h depends on what beliefs are contained in ES, which is usually an empirical issue, especially if we are referring to the ES of some particular person or group.

9. The Problem

We are now in a position to state the problem referred to in the title of this chapter. Suppose you accept the idea that there are in use objective, empirical concepts of evidence of the sort I have been describing (potential, veridical, and ES). How are the definitions of these concepts to be applied to actual cases to determine whether some particular e is in fact evidence that some particular h is true? How, for example, are we to determine whether, for such an e and h, the probability that there is an explanatory connection between h and e, given e, is greater than one-half? To draw an analogy, my dictionary defines a circle as "a plane figure bounded by a single curved

---

28. For more discussion of these options, see *The Book of Evidence*, pp. 156–57.

line every point of which is equally distant from the point at the center of the figure." That sounds like a good definition to me. But the dictionary does not say how to determine whether some actual figure satisfies this definition. This definition may suffice for purposes of proving theorems in geometry, which is all we might want to do (so-called pure geometry). But if we want to apply the definition and theorems to actual physical figures (applied geometry), we will need to know how to determine whether something is a plane, and a point, and whether two points on the figure are equidistant from the point that is the center. Now, the objection might be raised, in developing a theory of evidence, one should have in mind more than providing some abstract definition that satisfies one's favorite principles. One should be able to say how to apply the definition to actual cases to determine whether e is evidence that h in such cases. And this I have not done.

In *The Book of Evidence*, I introduced what I called "the Dean's Challenge." A former dean of mine, who was a physicist critical of philosophy of science, complained that philosophers of science generally, and the present author in particular, produce theories about science that have little if any relevance for scientists. I can hear him shouting from the heavens (or wherever he is now): "See, you have done it again! How are we supposed to apply your theory to real cases?"

Philip Kitcher, in an essay on my work on evidence, begins with the "Dean's Challenge," and imagining what the dean would say about my work, writes as follows:

> "Peter," he says, "this is all very clever. You are very good at thinking up examples to show what is wrong with particular proposals, and you show that your own suggestions—that evidence requires a probability greater than half of an explanatory

connection between hypothesis and evidence—survives the various cases you have used to test other ideas. But how do we scientists figure out these probability judgments?"[29]

That is, how are my definitions of evidence supposed to be applied to real cases? If I had developed an a priori concept of evidence, the general sort of answer would be clear: for some e and h, it is an a priori question whether e is evidence that h; so to determine whether it is requires some type of a priori "calculation." For example, on the simple h-d view, we need to determine whether h entails e. On Hempel's satisfaction view, we need to determine whether h is "satisfied" by the individuals mentioned in e. On Carnap's objective Bayesian view, we need to express e and h in some formal language and then determine whether the probability of h, given e, is greater than the probability of h (which for Carnap is an a priori calculation). To be sure, some of these calculations may be difficult, but at least we know in principle what needs to be done and how to do it. But with an empirical concept of evidence the task is quite different. We need to determine the truth of empirical claims of the form "e is evidence that h." And my theory doesn't say how to do this, only what it means to say that a statement of this form is true.

In the two chapters that follow I will turn to a potential solution in terms of what is called a "scientific method." A scientific method generally consists of rules for determining how to make nondeductive inferences from observed phenomena to more general propositions. If such inferences conform to these rules, then we might conclude that the observed phenomena constitute evidence that the propositions inferred are true. And the rules might provide a

---

29. Philip Kitcher, "On the Very Idea of a Theory of Evidence," in Morgan, *Philosophy of Science Matters*, p. 85.

way of determining whether statements of the form "*e* is evidence that *h*," construed in accordance with my definitions, are true. This is the path I will explore. I will do so with respect to a famous set of methodological rules proposed by Isaac Newton in the third book of his *Principia*. I will provide an interpretation of those rules that is different from the way they are usually construed, and then show what happens when, understanding evidence in the ways I have proposed, we apply those rules to the case in which Newton himself applies the rules, viz. the law of gravity. Do these rules help us solve the "problem of evidence"?

Portrait of Sir Isaac Newton by Godfrey Kneller © Corbis.

Chapter 2

Newton's Rules

1. Newton's "Rules for the Study of Natural Philosophy"

Isaac Newton begins the third book of his *Principia* with a set of four general methodological rules.[1] They are explicitly used in generating the universal law of gravity from six propositions he calls "phenomena" together with laws and theorems which he has established earlier in the *Principia*. Here are Newton's Rules:

1. No more causes of things should be admitted than are both true and sufficient to explain their phenomena.
2. Therefore, the causes assigned to natural effects of the same kind must be, so far as possible, the same.
3. Those qualities of bodies that cannot be intended and remitted [qualities that cannot be increased or diminished—eds.] and that belong to all bodies on which experiments can be made should be taken as qualities of all bodies universally.
4. In experimental philosophy, propositions gathered from phenomena by induction should be considered either exactly or very nearly true notwithstanding any contrary hypotheses,

---

1. Isaac Newton, *The Principia*, trans. I. Bernard Cohen and Anne Whitman (Berkeley: University of California Press, 1999), pp. 794–96.

until yet other phenomena make such propositions either more exact or liable to exception.

Rule 1. This rule, called the "vera causa" (true cause) rule, involves at least a principle of simplicity, urging scientists to admit no more causes than are necessary to explain the phenomena. If one cause will suffice, then two are one too many. In commenting on this rule Newton writes: "For nature is simple and does not indulge in the luxury of superfluous causes." The rule is important in Newton's argument for his law of gravity, since he wants to show that one force of gravity operating in the universe will suffice to explain the phenomena with which he begins; multiple forces are not needed.

Another important component of the rule is the idea of a *true* cause. Newton does not explain what he means by this, and various interpretations are possible. I'll mention three. First, some scientists in the eighteenth and nineteenth centuries, either following or rejecting Newton's rules, took it to mean that scientists should infer causes of given phenomena only if causes of those types are already known to produce other phenomena. No new types should be admitted.[2] Newton never says that explicitly, nor should he, since the idea of "new" or "old" types of causes is vague. (Is gravity an old type of cause that had been observed, since it is a force that was known to produce accelerations of unsupported bodies near the earth? Or, because it is universal [holding for all bodies], is it a new type?)

---

2. Thomas Reid, for example, suggests such an understanding of the rule (*An Inquiry into the Human Mind on the Principles of Common Sense* [Edinburgh: Edinburgh University Press, 2000]). Commenting on this interpretation, William Whewell writes: "Now the Rule thus interpreted is, I conceive, an injurious limitation of the field of induction. For it forbids us to look for a cause, except among the causes with which we are already familiar" (*The Philosophy of the Inductive Sciences, Founded upon their History* [London: John W. Parker, 1840]). Chapter 13 on Newton's rules is reprinted in my *Science Rules* (Baltimore: Johns Hopkins University Press, 2004), quotation on p. 114.

Second, in view of Newton's insistence at the end of the *Principia* that no hypotheses be introduced (i.e., propositions not "deduced from phenomena"), perhaps this rule should be understood as saying that no causes of phenomena should be introduced unless it is experimentally established, at least subsequently, that those causes exist and produce those phenomena.[3]

Third, although again he does not say so, perhaps Newton's use of "true" in the rule should be construed as a commitment to scientific realism: scientists should aim to infer causes that really exist, not simply ones that "save the phenomena." This interpretation accords with his Rule 4, in which he claims that propositions inferred inductively from the phenomena "should be considered either exactly or very nearly true."

There is something puzzling about Newton's remark quoted above that "nature is simple and does not indulge in the luxury of superfluous causes." If this is a justification of Rule 1, then how does Newton know that "nature is simple?" Is this an empirical assumption? If so, is it to be justified by appeal to more specific propositions about simple causes that have been inferred using Rule 1? But that would be circular, even if we used Rule 3 to make the induction from other more specific causes. Is "nature is simple" a metaphysical assumption, not to be justified empirically? But at the end of the *Principia* Newton explicitly excludes from "experimental philosophy" metaphysical and other principles that cannot be "deduced from the phenomena" using his four rules. In section 4, I will present a way of interpreting Newton's rules that avoids these puzzling issues and accords with important parts (though not all parts) of his fundamental empiricism as well as his practice. This interpretation, I will argue, makes the rules very plausible.

---

3. William L. Harper suggests something like this in his important work *Isaac Newton's Scientific Method* (Oxford, UK: Oxford University Press, 2011), p. 170.

Finally, in this rule Newton uses the term "phenomena," which he does not define in the *Principia*. In an unpublished definition intended for the second edition he says that a phenomenon is "whatever can be seen and is perceptible ... either things external which become known by the five senses, or things internal which we contemplate in our minds by thinking."[4] He offers these examples: fire is hot, water is wet, gold is heavy, sun is light, I am, I think. In the *Principia* he lists six phenomena which are Kepler's second and third laws of motion applied to the motions of the planets and their satellites.[5] Judging from the definition and the unpublished and published examples, for Newton "phenomena" are facts that have been established by observation by members of the relevant community (in the *Principia*, the scientific community). They are not disputed, but are facts the community in general accepts, or at least would accept once the results have been made known.[6] They may be quite simple ones (e.g., "fire is hot") established by direct observation, or more complex ones (e.g., "the moon, by a radius drawn to the center of the earth, describes areas proportional to the times"—Newton's sixth "phenomenon"), established by inferences from other phenomena.

Newton starts his argument for the law of gravity with six such "phenomena." But, by contrast to some other writers, he does not consider these phenomena to be indubitable—either empirical or a priori certainties, not subject to challenge. Indeed, some of them (including, among them, ones he cites in the derivation of the law of

---

4. From a manuscript sheet translated by J. E. McGuire, "Body and Void in Newton's *De Mundi Systemate:* Some New Sources," *Archive for History of Exact Sciences* 3 (1966): 238–39.
5. Kepler's second law is that for each planet in its orbit about the sun a line drawn from the center of the sun to the planet sweeps out equal areas in equal times. The third law is that the square of the period of revolution of a planet is proportional to the cube of its distance from the sun.
6. For example, following Phenomenon 1, Newton writes: "Astronomers agree that their periodic times [the moons of Jupiter] are as the 3/2 power of the semidiameters of their orbits..." (*Principia*, p. 797).

gravity) are approximations, or as he himself would say, "very nearly true." They are justified by appeal to observations, they are accepted by the community in question, and if you have some empirical reason to doubt their truth, you need to produce that reason. Unlike Descartes before him, and positivists much later, Newton is not attempting to provide an epistemic foundation for knowledge by deriving all scientific propositions from some indubitable base, whether Cartesian intuitions or sense-data. You start not with these but with a limited set of observationally established facts accepted by the community that, you hope, will lead you to establish a causal law that will explain those facts.

Rule 2. According to Newton, the second rule follows from the first. ("Therefore, the causes assigned to natural effects of the same kind must be, so far as possible, the same.") The idea is that if a cause is assigned to an effect of a given kind, then if you assign different causes to other effects of that kind you are violating the first rule; you are introducing "superfluous causes." Presupposed by this rule, or at least the way Newton treats it in his argument for gravity, is the idea that you can infer causes from effects, and the idea that if effects are the same, then you can infer that the cause of these effects is the same. If you have motions of the same kind in various planets (e.g., that "the planets . . ., by radii drawn to the sun, traverse areas proportional to the times"—Newton's fifth "phenomenon"), then you can infer that the cause of these motions is the same in each case. The rule, by itself, however, does not enable you to infer what that cause is—only that there is one, and it is the same in all such cases. Later I will comment on the important "so far as possible" clause.

Rule 3. "Those qualities of bodies that cannot be intended and remitted and that belong to all bodies on which experiments can be made should be taken as qualities of all bodies universally." This is an inductive rule insofar as it permits you to infer something about an entire class from characteristics of a portion of that class. But it is different from more standard formulations of induction, such as the following from Mill: "Induction is the process by which we conclude that what is true of certain individuals of a class is true of the whole class."[7] For Mill, induction is a certain type of inference—one that may be good or bad depending on the case. (Mill gives examples of both.) Newton's Rule 3 doesn't seem to allow for "bad" inferences. Perhaps this is because the rule itself is more restrictive than Mill's. It is restricted to "bodies," and to "qualities of those bodies that cannot be intended and remitted."

Cohen and Whitman, translators of the latest English edition, explicate the "intension and remission" exclusion as "qualities that cannot be increased and diminished."[8] In his commentary, Cohen gives examples: "Thus extension (the property of occupying space) and inertia (i.e., force of inertia or mass) are properties of this sort."[9] As Cohen puts it, they "do not vary." The "intension and remission" clause is both vague and controversial. If Cohen and Whitman are right, then extension and mass are qualities of bodies that cannot be "intended or remitted." But notice that Cohen understands these as the quality of occupying space and the quality of having mass (and in the important case of gravity, as the quality of exerting a gravitational force). The problem is that extension, mass, and gravitational force do "vary" in the sense that different amounts are possible. That is why Cohen formulates these qualities as "the quality of having x," (i.e., of having some amount or degree of x).[10]

If that is what Newton had in mind, then the clause in question is much less restrictive than Cohen seems to think, as Ernan

---

7. John Stuart Mill, *A System of Logic* (London: Longmans, 1872), p. 188.
8. *Principia*, p. 795.
9. *Principia*, p. 199.
10. This is, or is close to, an interpretation proposed by Steffen Ducheyne in his *The Main Business of Natural Philosophy: Isaac Newton's Natural-Philosophical Methodology* (Brussels: Springer, 2012), pp. 115–18. He interprets the clause as referring to "qualities or forces that can[not] be lost or taken away" (p. 117).

McMullen points out in an important commentary.[11] That is because the same could be said for color and heat, which Newton explicitly classifies as being subject to "intension and remission." In addition, the clause, so interpreted, would seem to lead to a result that Newton himself would reject. Consider the quality Q of being a body on which experiments can be made. Q, assuming it is a quality, seems to satisfy the Cohen-Whitman idea of being one that cannot be intended and remitted. Now Q belongs to all bodies on which experiments can be made. Therefore, by Rule 3, Q belongs to all bodies universally. So we may conclude that all bodies are bodies on which experiments can be made, which is not at all what Newton wants to say.[12] McMullen concludes that Newton "could have omitted the troublesome intensity criterion from the published version of Rule 3, without in the least affecting the manner of applying the Rule to concrete cases."[13] (I will return to this later when I provide my own proposals for how best to understand Newton's rules.)

There is another respect in which Newton's Rule 3 differs from some more standard versions of induction, which characterize induction as involving an inference from *observed* members of a class to all members. (Mill's definition above doesn't do that, but others do, and it is clear from his discussion that these are the ones Mill has in mind.) By contrast, Newton's Rule 3 characterizes the inference as one from a quality belonging "to all bodies on which experiments can be made" to the conclusion that these are "qualities of all bodies universally." Cohen, following Mandelbaum,[14] refers to this as "transdiction"—an inference

---

11. Ernan McMullen, *Newton on Matter and Activity* (South Bend, IN: Notre Dame Press, 1978).
12. Thanks to Marc Lange for this sort of example. In section 5 below, after presenting my own views on Newtonian induction, I will indicate why such an example is precluded (see n. 40). But Newton's Rule 3, as formulated, does seem to be subject to this problem.
13. McMullen, *Newton on Matter and Activity*, pp. 12–13.
14. Maurice Mandelbaum, *Philosophy, Science and Sense Perception* (Baltimore: Johns Hopkins Press, 1964), 74ff.

from all observable bodies to all unobservable ones. The problem with such an interpretation is that Newton cites inferences governed by Rule 3 that are not transdictions in this sense. For example: "That all bodies are movable and persevere in motion or in rest by means of certain forces (which we call forces of inertia), we infer from finding these properties in the bodies that we have seen." In this case, we are inferring that all bodies have a certain property P from the fact that all *observed* bodies have P, even if there is an intermediate step from "all observed bodies have P" to "all observable bodies have P." I will consider Rule 3 to apply to both sorts of inferences. What Cohen and Mandelbaum want to emphasize is that Newton uses his Rule 3 to allow inferences to unobservables. We can get that result simply be construing Rule 3 to allow an inference from what qualities observed bodies have to the claim that all bodies, whether observable or not, have those qualities.

As he does in his discussion of Rule 1, Newton appeals to an assumption about nature in defense of the rule: "nature is always simple and ever consonant with itself." Later I will deal with this claim.

Rule 4. "In experimental philosophy, propositions gathered from phenomena by induction should be considered either exactly or very nearly true notwithstanding any contrary hypotheses, until yet other phenomena make such propositions either more exact or liable to exceptions." This rule contains four important ideas. First, inductive conclusions from phenomena (presumably Newton has in mind inferences satisfying Rule 3) should be considered to be true, whether exactly or approximately. This, in conjunction with Rule 1, seems to reflect Newton's scientific realism, or at least some central ideas of that doctrine, according to which scientists should aim to arrive at propositions (including ones about unobservables) that are true, not simply ones that "save the phenomena" but could be false; and that using causal and inductive reasoning they can arrive at conclusions that can justifiably be regarded as true.

Second, the force of an inductive conclusion is not blunted simply by imagining a contrary hypothesis to explain the phenomena. Here Newton probably has in mind Descartes, who concocts imaginary contrary hypotheses in order to cast doubt on the one that we have accepted.

Third, the only way to produce exceptions to the inductive conclusion, or, presumably, to refute it,[15] is to discover new phenomena that will have such effects. Using his inductive Rule 4, together with the others, Newton arrives at his law of gravity involving one universal inverse-square force. This law is not weakened or refuted simply by imagining the contrary hypothesis that the law holds only for our solar system and nowhere else. Only new phenomena supporting this contrary hypothesis can have that effect.

Fourth, in saying that an inductive conclusion may be considered to be "very nearly true," and that new phenomena may make the conclusion "either more exact or liable to exceptions," Newton is recognizing something that modern physicists take for granted. The inductive conclusion may be an idealization, or it may hold exactly or approximately only under certain conditions. Newton commentators George E. Smith[16] and William L. Harper[17] emphasize the "idealization" and "making more exact" aspect of Newton's method. Even if the inductive conclusion itself is considered "exactly true," the argument leading to it may contain propositions that are idealizations. Kepler's laws on which the argument is based do not hold exactly because of perturbations produced by other planets. These perturbations can themselves be explained by invoking the law of gravity.

---

15. Newton here does not speak of refuting; I will take the liberty of doing so.
16. George E. Smith, "The Methodology of the *Principia*," in I. Bernard Cohen and George E. Smith, eds., *The Cambridge Companion to Newton* (Cambridge, UK: Cambridge University Press, 2002).
17. Harper, *Isaac Newton's Scientific Method*.

In the *General Scholium* at the end of the *Principia*, Newton adds some important methodological remarks to supplement his four rules. He defines a "hypothesis" as a proposition "not deduced from the phenomena," by which he seems to mean one that is not derived from phenomena in accordance with his four rules, and not derived from other propositions that have been "deduced from the phenomena." He then goes on to make the very strong claim that hypotheses "have no place in experimental philosophy." The trouble here is that Newton does introduce even what he calls "hypotheses" in his physics. (For example, "Hypothesis 1" in Book 3, p. 816, states that "the center of the system of the world is at rest.") Later I shall try to remove this contradiction. Also, in chapter 3, I will introduce an extension of this methodology, which Newton called the "method of synthesis." But first, I will reconstruct Newton's general argument for his law of gravity that uses his four methodological rules.

2. Newton's Argument for the Law of Gravity

The argument is complex. It begins with the six "Phenomena" and continues until Proposition 7 ("Gravity exists in all bodies universally and is proportional to the quantity of matter [mass] in each"), and Corollary 2 to that proposition ("The gravitation toward each of the individual equal particles of a body is inversely as the square of the distance of places from these particles")—some fifteen pages in the Cohen-Whitman translation. Here I will present a summary of that argument—just enough to say, in the sections that follow, how I think Newton is, or at least ought to be, using his four "rules for the study of natural philosophy," and how those rules might be construed to yield a plausible methodology.

First, Newton introduces six "Phenomena." These are propositions stating that the observed motions of the planets and their satellites satisfy Kepler's second and third laws of planetary motion.[18] For example, Phenomenon 1 states:

> The circumjovial planets [or satellites of Jupiter], by radii drawn to the center of Jupiter, describe areas proportional to the times (Kepler's 2nd law applied to the satellites of Jupiter), and their periodic times—the fixed stars being at rest—are as the 3/2 powers of their distances from the center (Kepler's third law).

In a more modern expression, for each satellite of Jupiter a line from the center of Jupiter to that satellite sweeps out equal areas in equal times, and for each satellite the square of the period of its revolutions around Jupiter is proportional to the cube of its distance from Jupiter. Phenomenon 2 repeats the claim for the satellites of Saturn. Phenomena 3, 4, and 5, if considered together, repeat these ideas for the five planets known to Newton (Mercury, Venus, Mars, Jupiter, and Saturn) with respect to the sun. Phenomenon 6 says that a line drawn from the center of the earth to our moon, in its orbit around the earth, sweeps out equal areas in equal times.

---

18. As Harper notes (*Isaac Newton's Scientific Method*, p. 139), Newton does not explicitly cite Kepler's first law as one of the Phenomena (that the planets revolve around the sun in elliptical orbits with the sun at one of the foci). First, the orbits are not exactly elliptical (because of perturbations from other planets); second, the sun is not at a focus; third, Newton wants to derive elliptical orbits as an idealization, assuming that the only force acting on each planet is the gravitational force exerted by the sun. Many-body problems are just too difficult.

We can summarize these Phenomena by writing the following as the first step in the argument:

> *Step 1*: All the observed planets and their satellites obey Kepler's second and third laws of planetary motion (in their orbits a line drawn to them from the orbited body sweeps out equal areas in equal times, and the square of their periods of revolution is proportional to the cube of their respective distances from the orbited body).

Next, Newton presents a series of statements that he calls both "Propositions" and "Theorems." Proposition (Theorem) 1 states:

> The forces by which the circumjovial planets [or satellites of Jupiter] are continually drawn away from rectilinear motions and are maintained in their respective orbits are directed to the center of Jupiter and are inversely as the squares of the distances of their places from that center.

This he derives from Phenomenon 1 (above) together with various propositions he has proved in Book 1 of the *Principia*. The general idea behind the latter is that if a body revolves around another, it is subject to a force (since it is changing direction) directed toward the stationary body (a centripetal force). He shows that the areas described by radii drawn to the central body are proportional to the times in which they are described, and also that if the periods are as the 3/2 power of the radii (i.e., that the square of the period is proportional to the cube of the radius), then the centripetal force will be inversely proportional to the square of the radius.

In Proposition (Theorem) 2, Newton repeats the ideas in Proposition 1, but now for the planets with respect to the sun, and in

Proposition 3, for the moon with respect to the earth.[19] Proposition 4 states that "The moon gravitates toward the earth and by the force of gravity is always drawn back from rectilinear motion and kept in its orbit." In his discussion of this proposition, on the basis of empirical calculations, he argues that the centripetal force by which the moon is kept in its orbit around the earth is the same in magnitude as the force of gravity on the earth—that is, the force causing bodies to fall toward the earth. He then argues, in accordance with his methodological Rules 1 and 2 ("vera causa," and "similar effects, same cause"), that these forces are not two but one and the same force. He defends this claim by supposing that the earth had several moons, the nearest of which "nearly touched the tops of the highest mountains on earth." By Newton's empirical calculations the centripetal force acting on such a moon would

> cause this little moon, if it were deprived of all the motion with which it proceeds in its orbit, to descend to the earth—as a result of the absence of the centrifugal force with which it had remained in its orbit—and to do so with the same velocity with which heavy bodies fall on the tops of those mountains, because the forces with which they descend are equal. And if the force by which the lowest little moon descends were different from gravity and that little moon were heavy toward the earth in the manner of bodies on the tops of mountains, this little moon would descend twice as fast by both forces acting together. Therefore, since both forces—namely, those of heavy bodies and those of the moons—are directed toward the center of the

---

19. For the proposition regarding the moon, Newton makes the simplifying assumption that the moon's apogee (the point in its orbit farthest from the earth) does not vary.

earth and are similar to each other and equal, they will (by [methodological] Rules 1 and 2) have the same cause. And therefore that force by which the moon is kept in its orbit is the very one that we generally call gravity. For if this were not so, the little moon at the top of a mountain must either be lacking in gravity or else fall twice as fast as heavy bodies generally do. (p. 805)

Earlier I summarized Newton's "Phenomena" in step 1. Let me now summarize this moon argument as follows:

Step 2: In accordance with methodological Rules 1 and 2, since the centripetal force keeping the moon in its orbit and the force of gravity are both directed toward the earth and have equal magnitudes, we infer that they are the same force.

Proposition 5 asserts that the moons of Jupiter and of Saturn gravitate toward their respective planets, that the planets gravitate toward the sun, and that by the force of their gravity are drawn back from rectilinear motions and kept in their orbits. In the discussion of this proposition, Newton argues that the revolutions of all of these bodies "are phenomena of the same kind as the revolution of the moon about the earth, and therefore (by [methodological] Rule 2) depend on causes of the same kind." He expands on this in a Scholium following Proposition 5, saying that although earlier he was calling the forces "centripetal," he has now established that this force is the same force as gravity. He concludes: "For the cause of the centripetal force by which the moon is kept in its orbit ought to be extended to all the planets, by [methodological] Rules 1, 2, and 4." In short, we have one force, not many different ones, for all the planets (and satellites). So,

Step 3: In accordance with Rule 2, since the motions of the planets with respect to the sun, and their satellites with respect to their planets, is the same in the above respects as the motion of the moon with respect to the earth, we infer that these motions have the same cause, viz., an inverse-square force proportional to the masses of the bodies. In accordance with Rule 1, since one force suffices for all these phenomena, we should infer that this is the only force acting. And in accordance with Rule 4, we should assert the truth of this conclusion.

Proposition 6 asserts that all bodies gravitate toward each planet, and that at any given distance from the center of any planet the weight (i.e., the "heaviness") of any body toward that planet is proportional to the mass of that body. Newton's argument for the latter claim is an experimental one involving the use of pendulums. Corollary 2 to Proposition 6 states that "all bodies universally that are on or near the earth are heavy [or gravitate] toward the earth, and the weights of all bodies that are equally distant from the center of the earth are as the quantities of matter [masses] in them." And he concludes: "This is a quality of all bodies on which experiments can be performed and therefore by [methodological] Rule 3 is to be affirmed of all bodies universally."

Finally, in Proposition 7, Newton writes "Gravity exists in all bodies universally and is proportional to the quantity of matter in each." Although this proposition talks about all bodies, Newton's discussion of it that follows is confined to the planets and their parts. Again, the idea seems to be that by the inductive Rule 3, we can generalize to all bodies universally.

Summarizing this part of the argument, we might say,

EVIDENCE AND METHOD

Step 4: By methodological Rule 3 (induction), all bodies gravitate toward each planet, and, even more generally, all bodies gravitate toward all others, in proportion to the quantity of matter (mass) in each. And by methodological Rule 4, these inductions can be considered true or very nearly true, until other phenomena show the need to make the conclusions more precise, indicate exceptions (or, presumably, reject them).

3. What's Going on Here? Newton's "Strategy"

First, notice that in generating the conclusion in each of the four steps I have outlined, empirical assumptions are introduced by Newton as at least part of the defense of that conclusion. In step 1, the conclusion that each of the planets and their satellites obey Kepler's laws is justified by Newton by reference to data gathered from observations of various astronomers. For example, in defense of Phenomenon 1 (about the motion of Jupiter's moons), Newton cites data about the distances of the four moons of Jupiter from Jupiter gathered by astronomical observations of Borelli, Townly, and Cassini using different methods of observation. In step 2, the conclusion is that the centripetal force keeping the moon in its orbit and the force of gravity on the earth are the same force. This is defended, at least in part, by empirical calculations involving data about the distance of the moon from the earth, the periodic time of the moon, and the circumference of the earth, yielding a calculation of how fast the moon would fall toward the earth if it were deprived of its inertial motion. In step 3, to reach the conclusion that the motions of the planets, their moons, and our moon have the same cause, viz. an inverse square force proportional to the masses of the bodies, Newton depends on the "Phenomena," earlier propositions he has proved, and the moon argument—all empirically defended. (Earlier proved propositions are derived by Newton from his three laws of motion, which he regards as empirical since they are "deduced from the phenomena and are made general by induction."[20]) Finally, the conclusion in step 4 that all bodies gravitate toward all others is defended by appeal to empirically established propositions about the planets.

So far so good, readers of Newton may say. Of course, in inductive arguments of the sort Newton employs the premises are empirical. What do you expect? But in order to get from these empirical premises to the bold conclusions Newton inferred you need rules of inference, such as his four methodological rules. Although these are not empirical, they should be considered as important a part of the Newtonian argument as the empirical premises. For example, Newton can infer that all bodies gravitate toward all others not just because all planets gravitate toward all others but also because methodological Rule 3 operates in this case, allowing one to generalize from what is the case with the planets to all bodies whatever. After all, doesn't Newton himself use phrases like "by Rules 1, 2, and 4" in defending his inferences?

This interpretation of Newton's rules is fairly standard. For example, in his critique of Newton's rules, William Whewell in the mid-nineteenth century writes:

20. *Principia*, p. 943.

> In considering these Rules, we cannot help remarking, in the first place, that they are constructed with an intentional adaptation to the case with which Newton has to deal,—the induction of Universal Gravitation; and are intended to protect the reasonings before which they stand. Thus the first Rule is designed to strengthen the inference of gravitation from the celestial phenomena, by describing it as a *vera causa*, a true cause. . . . [T]he third rule appears intended to justify the assertion of gravitation, as a *universal* quality of bodies.[21]

Similarly, in a 2004 article on these rules, Quayshawn Spencer writes:

> Newton needs an inductive rule to instruct him that it is permissible to conclude from this similarity [lunar centripetal acceleration and terrestrial gravity] that these two phenomena are due to the same cause.[22]

Newton's rules are what allow you to infer the same cause from effects of the same kind; they are what allow you to generalize from a property present in observed (or observable) bodies to its being present in all bodies. Or, as Whewell puts it, they are intended to "protect," "strengthen," and "justify" such an inference. On this view, the rules are needed, in addition to the empirical premises, to justify, or help justify, an inference from these premises to the desired conclusions. It is because of the four rules that Newton can

---

21. Reprinted in Achinstein, *Science Rules*, pp. 113–14.
22. Quayshawn Spencer, "Do Newton's Rules of Reasoning Guarantee Truth . . . Must They?" *Studies in History and Philosophy of Science* 35 (2004): 759–82, p. 761. The same claim is made by George E. Smith, "Newton's *Philosophiae Naturalis Principia Mathematica,*" in *Stanford Encyclopedia of Philosophy* [online]). He speaks of the rules as "authorizing" inferences.

make the inferences he does.[23] Is this the only or the best way to interpret these rules?

Although Newton formulates the rules, and cites them in making various inferences, he does not explicitly address the question of how and when they are to be used. For example, in using Rule 2, can we make the inference from any two effects of the same kind, to the conclusion that their cause is the same? Even Newton qualifies this rule by saying "so far as possible." More important, what counts as "the same kind"?[24] Are two motions "of the same kind" if they both involve uniform velocity in a straight line? What if one motion involves no forces acting but the other is the resultant of opposing forces, such as a horse pulling a carriage with uniform velocity?[25] Are two effects to be counted as being of the same kind only if they have the same cause? If so, the rule is trivial. If not, are two effects of the same kind if some set of properties of these effects that we have observed are the same? Which set, and how many? And why suppose that we can infer the same cause from that set? After all, there

---

23. In a discussion of Newton's rules, William L. Harper sometimes seems to interpret Newton this way. For example, "The appeal to Rules 1 and 2, therefore, clearly counts against the alternative hypothesis that the centripetal force holding the moon in its orbit is a different kind of force" (*Isaac Newton's Scientific Method*, p. 172). And, "This rule [rule 3], therefore, endorses counting such parameter values found to be constant on all bodies within the reach of experiments as constant for all bodies" (p. 38). On other occasions (see below, n. 35) he may be interpreting Newton in a way similar to the one I will defend.
24. Whewell raises this question in objecting to Rule 2 (Achinstein, *Science Rules*, pp. 118–19).
25. Spencer claims that "without an inductive rule to permit an identity relation between the moon's *kind* of acceleration and the *kind* keeping the primary planets in orbit, Newton has only shown that these accelerations have the *same properties*. Invoking rules 1 and 2 classifies each of these accelerations as the same kind. Rules 1 and 2 are used in the same way to designate all celestial centripetal acceleration as the gravitational kind" ("Do Newton's Rules...," p. 762). My response is to say that it is not the rules that are doing the work here, but Newton, who is classifying these phenomena as being of the same kind, which is an empirical assumption. The rule itself doesn't tell Newton or anyone else when two phenomena are of the same kind, so that the same cause can be invoked.

are many effects that have a set of properties in common, yet we cannot infer the same cause. (Think of sets of symptoms of disease in medicine.) More generally, it is an empirical, not an a priori, question whether we can infer a common cause from effects that are the same in certain ways.

In the case of (the inductive) Rule 3, how are we to determine whether the qualities in question "belong to all bodies on which experiments can be made," other than by making prior inferences from "all bodies on which experiments *have been made* have those qualities"? How many such bodies need to be observed, and under what conditions, to make inferences to "all bodies on which experiments can be made" (which Newton himself does), or to "all bodies" (which he also does)? This seems to be an empirical question that depends on what bodies have been observed, the method of observation, and the types of qualities involved.

More generally, what is the justification for these rules? Are they a priori (like rules of deduction)? Are they rules that follow from broad metaphysical hypotheses about the simplicity of nature (as is suggested in some of Newton's remarks, even though Newton rejects metaphysical hypotheses in science)? Or are they rules that follow from empirical assumptions about nature? If so, without using the rules themselves, how would we know that nature operates in accordance with those rules?

Critics of Newton's methodology, such as William Whewell[26] in the nineteenth century, Paul Feyerabend[27] in the twentieth, and

---

26. Whewell, in Achinstein, *Science Rules*, pp. 112–23.
27. Paul Feyerabend, "On the Improvement of the Sciences and the Arts, and the Possible Identity of the Two," in R. S. Cohen and M. W. Wartofsky, eds., *Boston Studies in the Philosophy of Science*, vol. 3 (Dordrecht: D. Riedel, 1968), pp. 387–415.

John Norton[28] and Victor Di Fate,[29] in the twenty-first, pounce on these and other issues, declaring Newton's rules worthless as a means of testing or justifying inferences. Whewell and Feyerabend think that conforming to such rules will stifle science, since it will restrict science to known or established causes. Norton and Di Fate argue that the rules have no justificatory force whatever. I want to suggest another way of viewing them, according to which they do have justificatory force and no stifling effects. Whether this is Newton's own way of understanding the rules I hesitate to say, but I would be happy to attribute this to him, since I think this interpretation makes them plausible. Newton does not provide answers to questions of the sorts raised in the last few paragraphs; and in the absence of such answers, I will propose a way of understanding Newton's rules that conforms reasonably well with the use he makes of them in deriving the law of gravity as well as with his general strategy in the *Principia*.

In his Preface to the first edition, Newton writes about the aim of the work, as follows:

> For the basic problem of philosophy seems to be to discover the forces of nature from the phenomena of motions and then to demonstrate the other phenomena from these forces. It is to these ends that the general propositions in books 1 and 2 are directed, while in book 3 our explanation of the system of the

---

28. John Norton, "A Little Survey of Induction," in Peter Achinstein, ed., *Scientific Evidence* (Baltimore: Johns Hopkins University Press, 2005), pp. 9–34.
29. Victor Di Fate, "Achinstein's Newtonian Empiricism," in Gregory J. Morgan, ed., *Philosophy of Science Matters: The Philosophy of Peter Achinstein* (New York: Oxford University Press, 2011), pp. 44–58. Also, "Is Newton a 'Radical Empiricist' about Method?," *Studies in History and Philosophy of Science* 42 (2011): 28–36. See n. 41 below.

EVIDENCE AND METHOD

world illustrates these propositions. For in book 3, by means of propositions demonstrated mathematically in books 1 and 2, we derive from celestial phenomena the gravitational forces by which bodies tend toward the sun and toward the individual planets. Then the motions of the planets, the comets, and moon, and the sea are deduced from these forces by propositions that are also mathematical. (p. 382)

Given this, and given what he accomplishes in the third book, one of Newton's main aims in that book is this:

Aim: To establish a general law that invokes a cause (in this case, a force) to explain a range of phenomena that have been established by observation (in this case, Keplerian motions of bodies in the solar system).[30]

You have "established" a law if you have given sufficiently good reasons to believe it, and the law is true or approximately true. If the reasons you give consist of evidence, then to establish a law the evidence must be strong "veridical" evidence, in the sense defined in chapter 1. If, given your epistemic situation, you are justified in believing that you have strong veridical evidence, and hence that you have established the law, then your evidence will be ES-evidence, in the sense defined in chapter 1 (whether or not the evidence is veridical). For the purpose of achieving the aim above, an important part of what I will call Newton's "general strategy" is this:

---

30. I say that this is one of Newton's main aims because, as is indicated in the quote above, and as I will show in chapter 3 when discussing what Newton calls the method of synthesis, another of his aims is to use the general law he establishes from the Keplerian planetary motions to explain many other phenomena as well.

1. Start with observed phenomena (in this case, observed Keplerian motions of the planets and their satellites).
2. Infer a cause of these phenomena (in this case, a gravitational force), and if possible, a unique cause, from the phenomena themselves, making use of empirical propositions derived earlier.
3. Generalize this cause (gravitational force) to as large a class as possible (e.g., to all bodies) so that a general law can be formulated mathematically.
4. Regard the strategy outlined above as successful if in using it we arrive at a general causal law that can be accepted as true, or approximately true.

Whether this strategy can be carried out successfully in a given case is an *empirical* question for that case. It needs to be determined empirically, as Newton does, whether the six Phenomena he cites involving the motions of the planets and satellites are in fact the case or approximately the case. It needs to be determined empirically whether the causes of these motions are inverse-square centripetal forces exerted on the planets and their satellites (as Newton does by invoking earlier principles derived from his empirical laws of motion). It needs to be determined empirically whether these forces are identical (as Newton does in his moon argument). And it needs to be determined empirically whether the inverse-square law involving the planets and their satellites can be generalized to all bodies. If these empirical defenses can be given, and are good ones, then we can regard the general law that is derived using the strategy as true or approximately true, and hence as established. Whether it is (truly) established, in the sense I have indicated, depends on whether it is (approximately) true.

4. An Interpretation of Newton's Rules

In the light of this general strategy, how might Newton's four methodological rules be regarded? They are quite different from standard rules of deduction that tell you what you can always infer from what (e.g., from a sentence p and a sentence of the form "if p then q," you can always infer q). Nor are they like rules that some philosophers include in what they call inductive logic (e.g., from propositions of the form "90 percent of As are Bs," and "this is an A," you can always infer, with a probability of .9, "this is a B"[31]). Newton's Rule 1 does not say that if a cause is true and sufficient to explain the phenomena, then you can always infer that it exists and causes those phenomena. It says or implies only that if a cause is to be admitted it must be true and sufficient to explain the phenomena (i.e., if it is not true and sufficient to explain the phenomena, then don't infer it); and that if causes C_1 and C_2 are both true and sufficient to explain the phenomena, don't infer both. Rule 2 does not say that if natural effects are of the same kind, then you can or should always infer that they have the same cause, but only that "so far as possible" you should try to assign the same cause to the same effects. Only Rule 3 seems to conform to what philosophers usually regard as an inductive rule of inference, since it seems to allow—indeed, to require—inferences from "such and such qualities are qualities of all observed (or observable) bodies" to "they are qualities of all bodies." Rule 4, does not tell you what proposition to infer from what, but only that if you have followed Rule 3 (and perhaps the others), and have arrived at a proposition of the sort you seek, then you should infer that the proposition is true or approximately true.

---

31. In fact such a rule is invalid; it depends on how the particular A was selected from the class of As. (That's my point about inductive rules.)

These rules are to be understood as relativized to a fairly specific aim—that of establishing a law to explain a range of phenomena. If standard rules of deduction or induction are to be associated with some aim, the latter would have to be described in a much more general way—for example, truth, or inferring truths from truths, or rationality. Newton's rules are much more like rules of strategy than rules of inference. Rules of strategy also exist in deduction and induction (e.g., "In a deductive proof, try to use existential instantiation (EI) before universal instantiation (UI)"). They suggest steps you should take, or try to take, if you want to achieve your aim. Moreover, they are rules of strategy for how to proceed to *defend* a law, for what philosophers of science refer to as "justification," rather than "discovery," although, as I have argued elsewhere,[32] there is no incompatibility between using the rules in the process of discovering the idea of the law as well as in its defense.

I can offer no problem-free interpretation of the rules that incorporates (a) everything they explicitly say, (b) everything Newton says about them in his discussions of them, and (c) how Newton uses them in deriving the law of gravity. But I will offer a proposal that I believe yields rules that are reasonable and that reflects many of Newton's ideas. It involves regarding Newton's methodological rules as very general rules of strategy for achieving the aim of establishing a general law that invokes a cause to explain a range of phenomena. The first three of them tell you what to try to do if your aim is to establish a law of this sort. And, as I will indicate later, it can be shown that if you have this aim, and you follow the rules in an empirically defensible way, then, in accordance with the fourth rule, you can conclude that the law you aim to establish is true or approximately true.

32. Peter Achinstein, *Law and Explanation* (Oxford: Clarendon Press, 1971).

Here is the way I would understand these rules: if your aim is the one noted, then:

(rule 1)[33] When you introduce a cause or causes, try to empirically determine (a) whether the causes introduced exist (e.g., whether there are centripetal forces exerted by the sun on the planets); (b) whether these causes are different or the same (e.g. whether these centripetal forces are the same force or different ones); (c) whether the cause(s) postulated not only exist, but actually cause, or are part of the cause of, the phenomena in question (e.g., whether the Keplerian motions of the planets and their moons are in fact caused by the gravitational force postulated); (d) whether the cause(s) postulated is (are) sufficient to produce, and hence explain, the phenomena, or whether additional causes are required (e.g., whether the gravitational force postulated to explain the Keplerian motions is sufficient to explain those motions, or whether some additional force is required to explain them). Thus, (a), (b), and (c) are intended to be a strategic analogue of Newton's idea of a "true cause," (d) his idea of admitting a cause "sufficient to explain their phenomena."

(rule 2) When you introduce a cause or causes for a given effect, try ("so far as possible") to assign those causes to other effects of the same kind. This involves trying to determine empirically whether such an assignment is reasonable (I would interpret "so far as possible" to mean "so far as empirically possible"). For example, if you have empirically established that the fact that bodies fall at a certain rate of acceleration toward the earth is caused by an inverse-square force of attraction exerted by the earth, try to determine how a moon orbiting close to the earth would fall if it suddenly lost its

---

33. In what follows in presenting and referring to my formulation of these rules, I will use a small *r*, and when referring to Newton's formulation, a capital *R*.

inertial motion—in order to see whether the same cause could be assigned to both motions. This idea is intended to represent a strategic counterpart to Newton's idea that "nature is simple and does not indulge in the luxury of superfluous causes." However, we do not assume as a grand metaphysical or empirical hypothesis that nature is simple. We try to find out empirically in a given case what number of different causes and laws are needed.

(rule 3) If you have established that some property holds for all observed (or observable) bodies, try to generalize this to all bodies, or, if exceptions be found, to all bodies in some restricted class. The more general you can make the claim, the better.[34] This idea is intended to reflect (at least part of) Newton's requirement that established propositions (e.g., that the observed planets obey the law of gravity) should be "made general by induction."

(rule 4) If you have followed rules 1–3 in an empirically defensible way, and have arrived at a causal law, then you may infer that the causal law is true or approximately true; you are justified in doing so until phenomena are discovered that cast doubt on the law as it stands. If such phenomena are discovered, you should try to restrict the law to situations in which it holds more exactly, or (my addition) if too many such restrictions are necessary, you should reject the law. You should not regard any contrary hypothesis that explains the phenomena as casting any doubt on the causal law, unless that contrary hypothesis is arrived at by following rules 1–3 in an empirically defensible way.

---

34. This is stressed in Newton's *Opticks* when he writes: "And although the arguing from Experiments and Observations by Induction be no Demonstration of general Conclusions; yet it is the best way of arguing which the Nature of Things admits of, and may be looked upon as so much the stronger, by how much the Induction is more general . . ." (New York: Dover, 1979), pp. 404–45.

EVIDENCE AND METHOD

So understood, the rules tell you in a very general and abstract manner what you should try to do if your aim is to establish a causal law to explain a range of phenomena.[35] They are not based on metaphysical or empirical assumptions about the nature of the world (e.g., the assumption that nature is simple), as Newton seems to suggest. They do not assume that there is a single cause that will explain the phenomena. They do not even assume that all phenomena have causes, or that there are laws that govern the phenomena in question. And even if you assume that the phenomena you have observed have causes subject to general laws, the rules are silent as to what these causes and laws might be. They do not require you to search for, or infer, causes of the phenomena in question that are among types known to operate in the case of other phenomena.

If you follow these rules, will you achieve the aim of establishing a general law? Not necessarily. The law might turn out to be false (as was indeed the case with Newton's law of gravity, e.g., when bodies approach the speed of light). Will you at least be justified in believing that the law arrived at is true, or approximately true? Not necessarily if you follow just rules 1–3, which require only that you *try* to do certain things. On the basis of these three rules, you could try to give empirical grounds justifying the cause and law arrived at, even if those grounds are insufficient for doing so. It depends on the empirical facts you cite or have in mind. Are phenomena you are investigating really of the "same kind" so as to allow an inference to

---

35. There are passages in Harper that suggest such an interpretation of the rules. For example: "This rule [Rule 1] is, explicitly, formulated as a guide for inferring causes of natural things from their phenomena" (*Isaac Newton's Scientific Method*, p. 170). Whether his view is the same as the one I am developing depends on whether he believes that these "guides" for inferring causes have epistemic force— i.e., whether at least part of the reason you are justified in believing the conclusion of the inference is that you have followed the guide. If he believes this, his view is different from mine.

a common cause? Are the phenomena you are investigating sufficiently numerous and varied to allow any generalization of your results, or one that is as broad as you have made it? These are empirical questions not settled by rules 1–3. If, as I have done above, you add rule 4, with its requirement that you follow rules 1–3 in an empirically defensible way, then you will be justified in believing that the law arrived at is (approximately) true. But whether you have followed rules 1–3 in an empirically defensible way needs to be established independently of the rules (as Newton himself does in the moon test argument for inferring a single force for the orbiting moon and falling bodies).

The big catch here, of course, is the requirement in rule 4 that you follow rules 1–3 in a way that is "empirically defensible." If we understand this in a strong sense, it is a requirement that is not relativized to a scientist or an epistemic situation. In this sense, if, for example, following rule 3 you have made an inductive generalization to a causal law, then whether your generalization is empirically defensible does not depend on you or your epistemic situation but only on the particular empirical facts of the case. (A nonrelativized analogue in mathematics would be reasoning that is "mathematically defensible," except, of course, this would be establishable a priori.) There is also a sense of "empirically defensible" that is relativized to an epistemic situation. This is the one I take to be most relevant for the idea of justification employed in rule 4. Whether one is justified in inferring that a law is true depends on one's set of beliefs. Newton was justified in inferring his law of gravity, given his epistemic situation, but we are not, given ours. Whether Newton was so justified—whether he followed the rules in an empirically defensible way—depends on empirical facts about what he believed and why, which cannot be determined using the rules themselves.

So, critics will ask (and have asked) what's the point of having such rules? Of what value are they? To this question I now turn.

5. Of What Value Are These Rules?

In dealing with this question, I want to start by contrasting Newton's strategy and his rules, as I have formulated them above, with two others that he rejected, Cartesian rationalism and hypothetico-deductivism. Both present methods and rules for arriving at and establishing laws. These methods are intended to be such that we can determine a priori whether they have been successfully followed, and hence whether we are justified in believing the law arrived at.

Descartes presents twenty-one rules in his posthumously published book *Rules for the Direction of the Mind*. For our purposes we can concentrate on Rule 3:

> *Descartes' Rule 3*: Concerning objects proposed for study, we ought to investigate what we can clearly and evidently intuit or deduce with certainty, and not what other people have thought or what we ourselves conjecture. For knowledge can be attained in no other way.[36]

On Descartes' view, we try to derive everything in science from what he calls "intuitions," which are characterized as "the indubitable conception[s] of a clear and attentive mind which proceeds solely from the light of reason." These are a priori thoughts that

---

36. Reprinted in Achinstein, *Science Rules*, p. 19.

are immediately evident just by thinking them. His examples include mathematical ones, such as "2 + 2 = 4," as well as ones about the physical world, such as "bodies are extended." From the intuitions we derive conclusions by what he calls "deductions," which are inferences to propositions that follow necessarily from the intuitions, where the inferences involve a continuous, uninterrupted train of thought. A mathematical example would be an inference to a theorem in geometry from axioms that are intuitions. A physical example would be Descartes' inference in physics to the law of inertia, which (according to him) follows from intuitions about God and his simplicity. According to Descartes, we establish the law of inertia by starting with intuitions about God, his creation of matter and motion, and his simple mode of operation, deducing (in his sense) the law of conservation of motion (that the total amount of motion in the universe remains constant), and from this, deducing that "a moving body, so far as it can, goes on moving" (part of Descartes' first law of motion).

Critics of Descartes are quick to point out that his concept of an "intuition" is unclear. But Newton the empiricist objects to this method by claiming that in physics those propositions Descartes claims to know by pure thought cannot be so known. In his discussion of his own Rule 3, Newton writes:

> The extension of bodies is known to us only through our senses.... That all bodies are impenetrable we gather not by reason but by our senses.... That all bodies are movable and persevere in motion or in rest by means of certain forces (which we call forces of inertia) we infer from finding these properties in the bodies that we have seen [and not from intuitions about God]. (p. 795)

If you are an empiricist about the physical world, you will reject Descartes' methodology and agree with Newton that knowledge of the physical world is not a priori, but can be gained only by appeal to facts established empirically.

A second methodology Newton rejects is the method of hypothesis, or hypothetico-deductivism, the basic form of which is this: You start with observed phenomena that you want to explain (e.g., Newton's six Phenomena regarding the Keplerian motions of the planets and their satellites). You then introduce a hypothesis not yet experimentally established, and you determine whether from this hypothesis, together possibly with other assumptions, you can deductively derive, and thereby explain, the phenomena you began with, as well as new phenomena that your hypothesis predicts. If you can, then your hypothesis is established, depending on whether the new as well as the old phenomena derived in fact obtain.

Newton rejects this methodology. In a letter to Oldenburg of June 2, 1672, Newton writes:

> If any one offers conjectures about the truth of things from the mere possibility of hypotheses, I do not see how anything certain can be determined in any science; for it is always possible to contrive hypotheses, one after another, which are found rich in new tribulations.[37]

This suggests what later became known as the "competing hypothesis" objection. There may be another hypothesis, incompatible with yours, from which the same phenomena can be derived and

---

37. *Principia*, appendix, p. 673.

explained.[38] If so, then, according to the hypothetico-deductivist, that hypothesis would also be established. But two incompatible hypotheses can't both be established on the basis of the same phenomena. From the "mere possibility" of a hypothesis—that is, from the mere fact that it entails and explains a range of phenomena—you cannot conclude that it is true.

Earlier I noted Newton's famous claim that "hypotheses have no place in experimental philosophy." Yet he introduces propositions in the *Principia* that he explicitly labels "hypotheses." Can his view on hypotheses be formulated so as to avoid this contradiction? My suggestion is to interpret him as saying that you can introduce hypotheses so long as you do not infer their truth from the fact that, if true, they could explain phenomena. In other words, you can do so, provided you don't draw the conclusions that hypothetico-deductivists do.

Both Cartesian rationalists and hypothetico-deductivists will claim a certain advantage over Newton's methodological rules as I have presented them. The advantage is that you are able to determine a priori whether your hypothesis is established if the phenomena explained and predicted have been established. For Descartes, whether a proposition is "intuited" (in his sense), or whether it is derived by "deduction" (in his sense) from intuitions, is a matter of pure a priori thought. In the hypothetico-deductive case, whether the phenomena obtain is an empirical matter. But whether

---

38. Mill develops this idea very explicitly in the mid-nineteenth century when he rejects the wave theory of light and particularly its hypothesis that an ether exists in which wave motion is produced. He writes: "Accordingly, most thinkers of any degree of sobriety allow that an hypothesis of this kind is not to be received as probably true because it accounts for all the known phenomena, since this is a condition sometimes fulfilled tolerably well by two conflicting hypotheses; while there are probably many others that are equally possible, but which, for want of anything analogous in our experience, our minds are unfitted to conceive" (*System of Logic*, p. 328).

EVIDENCE AND METHOD

the hypothesis entails those phenomena, and on the usual h-d view, whether if true it explains them, are not empirical issues but a priori ones, to be determined by pure reasoning. If you are sufficiently talented and sophisticated, you can do the "calculations" in your armchair. These methodologists will say that you can't do that with Newton's rules as I have interpreted them. They are not subject to any such a priori calculation to determine whether they have been successfully applied to establish a proposition. To be sure, calculations are crucial to Newton in arriving at propositions to which the rules can then be applied in making causal and inductive inferences. For example, Proposition 1 in Book 3, concerning the inverse-square character of the forces exerted on the moons of Jupiter, is derived mathematically from Phenomenon 1 and propositions that are theorems in Book 1. But whether an inductive generalization of this to all planets, and then to all bodies, is justified is not a mathematical calculation.

What can be said is that if you have followed rules 1–3 in an empirically defensible way, then you can conclude (by rule 4) that the general law you have inferred is (approximately) true. But whether you have followed rules 1–3 in an empirically defensible way is an empirical, not an a priori, matter. So if you reject Cartesian rationalism and hypothetico-deductivism, because they fail to give you justification for believing the laws you arrive at using them, and if Newton's rules will give you such justification, but only if they are applied in an empirically defensible way, are there any real advantages in using Newton's rules? Of what value are they?

One important value that Newton achieves by using the rules is as guideposts through a complex argument for the law of gravity. They can help you to see and understand the sorts of inferences Newton is making. As he does in his moon argument in Proposition 4, Newton can say that he is invoking methodological rules 1 and 2 to get you to see that he is making a causal inference that is in conformity with these rules. He is presenting empirical reasons to conclude that the inertial force keeping the moon in its orbit is the same force as gravity on the earth. He is doing so having shown empirically that these forces exist, that they are responsible for the motions in question, that they have the same values, and that if they were different, then the moon, if deprived of its inertial motion near the earth, would fall toward the earth twice as fast as bodies on earth do. By reference to the methodological rules you may be able to see what sorts of inferences Newton is making, and how he defends them. You will also be in a better position to assess those inferences, once they are made explicit, as Newton himself does. Few if any scientists follow Newton in identifying their causal and inductive inferences as such. Such inferences are empirical and nonmathematical, but crucial. Exposing them to full view is a virtue, not simply an exercise of the sort found in logic.

Do the rules provide any kind of justification for the inferences Newton makes? Let me distinguish two different senses in which a set of rules can be used to justify an inference. One I will call epistemic, the other, pragmatic. Suppose you make an inference from the propositions "this body is being subjected to an unbalanced force," and "if this body is being subjected to an unbalanced force, then it is accelerating," to the proposition "this body is accelerating." In deductive logic, the rule *Modus Ponens* tells you that such an inference is sound. This can be interpreted epistemically to imply that if you are justified in believing the first two propositions, then you are justified in believing the third. The inference is sound because it is in conformity with the rule.

Now contrast this with an inductive inference of the sort that Newton makes from the fact that all observed planets and their satellites obey the law of gravity to the claim that all bodies obey this law. You can't epistemically justify that inference by a rule that says that when all observed members of a class have a given property, then you may infer that all members of the class do (or you may infer this with high probability). That's a bad rule. It depends on what class you are speaking of, how many of its members you have observed, under what conditions, how varied your sample, and so forth. The same holds even if you restrict the rule to "bodies," as I have done in the previous section, and to "qualities of bodies that cannot be intended and remitted," as Newton does. Nor is the inference in question epistemically justified by Newton's rule 3 in my formulation. That rule only urges you to try to generalize inductively if you can, meaning if you are empirically warranted in doing so. Whether you are so warranted will depend on empirical facts about the class of bodies observed (how many, how varied, etc.). It is such facts that will give you epistemic justification for inferring the generalization, not the rule.[39]

39. Here I agree with Whewell, Norton, and Di Fate, each of whom claims that only the particular empirical facts, not the rules, give an epistemic justification for an inference. For example, in his criticism of Newton's Rule 3 (induction), Whewell writes: "The assertion of universality of any property of bodies must be grounded upon the reason of the case, and not upon any arbitrary maxim [such as Rule 3]" (Achinstein, *Science Rules*, p. 120). Norton asserts that "all induction is local," and is not driven by any universal rules (Achinstein, *Scientific Evidence*, p. 26). I can't resist quoting Norton from a personal correspondence: "Newton knows that there is a good case to be made for universal gravitation from the phenomena he lists. However, he is also aware that, in making the case, there are going to be some delicate moments at which an unkind critic could protest. (Newton, as we know, felt himself surrounded by unkind critics.) So what Newton does is to identify each of these moments of fragility and to concoct a general rule that can be summoned to carry past each weak moment. He writes them so as to have a general plausibility and then moves them to the front of the work. The idea is that a reader will read these rules, find them plausible in the abstract, assent to them and thereby be unable to dissent when Newton uses them later to pass the moments of awkwardness." All three of these authors conclude that Newton's rules have no justificatory force (a claim I will reject in a moment). Whewell and Norton conclude that the rules are indeed worthless (unless, according to Whewell, they are transformed into his version of "inference to the best explanation"). Di Fate sees their only role as that of indicating what sort of inference has been made, without justifying that inference (see n. 41).

More generally, I claim, there are no valid universal causal or inductive rules to epistemically justify particular causal or inductive inferences. Universal rules such as "from effects of the same kind you may infer the same cause," or "from the fact that all observed As are Bs, you may infer that all As are Bs" are not valid.[40] This is the case even if such rules are weakened by adding "usually" in front of "you may infer the same cause (or that all As are Bs)"; sometimes you can make such inferences, sometimes not. And when a causal or inductive inference is epistemically justified, it is so not in virtue of the validity of such rules, but in virtue of some empirical fact or set of facts about the character of the particular causes and effects or about the character of the properties in question and the sampling procedure. For an epistemic justification of causal and inductive inferences, no universal rules, or ones with "usually" included, are needed; particular (or general) empirical facts will do just fine for that purpose.

Nevertheless, rule 3, or rather an appeal to it, can provide an important type of nonepistemic justification, a pragmatic one, for making the inductive inference above. Newton's aim is to establish a general causal law that will explain a range of astronomical phenomena. And his rule 3, as I formulate it, urges you to try to generalize less general causal propositions (in Newton's terms, "make them general by induction"). Newton's inference is pragmatically justified by appeal to rule 3 because generalizing in accordance with the rule will get him to the type of causal law he seeks to establish empirically. It won't follow that he will be

---

40. This provides an answer to the simple type of counterexample by Marc Lange noted in section 1 (n. 12). From the fact that all observed bodies are observable you can't infer that all bodies are observable. There are no observational facts that support such an inference, and plenty that refute it.

epistemically justified in believing the law he obtains by such an inference. But he will be pragmatically justified in making the inference.[41]

There are various ways Newton's rules have been faulted, including a criticism of his appeal to a vague and unjustified assumption about the simplicity of nature, and his assumption (if indeed he makes it) that satisfying the rules by itself provides an epistemic justification for believing the conclusion that is drawn. My interpretation of the rules avoids both criticisms. Let me mention three other ways the rules have been, or might be, faulted:

1. By showing that following them cannot achieve the aim Newton seeks. (You might try to demonstrate that such an aim cannot be achieved at all, or, even if it can, that it is a worthless aim.)
2. By agreeing that the aim in using the rules is a good one and is achievable, but arguing that there are better strategies for achieving it that do not involve following these rules.
3. By claiming that the rules are trivial and uninformative. To say, for example, that you should try to generalize when you are empirically warranted in doing so is pretty much like saying in chess that you should try to checkmate your opponent when the conditions are appropriate.

---

41. As noted earlier, Di Fate rejects the justificatory role of Newton's rules. He writes: "what needs to be questioned, I submit, is whether the rules *are* intended to justify or defend inferences." Yet he adds: "a closer inspection of how they function in the argument of book 3 will reveal that they seem merely to *indicate* the inferences that Newton wants us to make, while the *grounds* or *defense* of those inferences consist in other empirical propositions that we have already been brought to accept by that point in the argument" ("Is Newton a 'Radical Empiricist,'" p. 31). My claim is that the appeal to the rules is a kind of defense of an inference, albeit a pragmatic nonepistemic one.

The first option is adopted by Cartesians who argue that establishing scientific laws means establishing them with absolute certainty, which for them requires the sort of a priori proof you find in mathematics, not causal and inductive reasoning of a sort Newton champions. This first option is also adopted by skeptics who believe either that there are no universal (or even statistical) causal laws or even if there are, they cannot be established by causal or inductive reasoning because, as Hume and Karl Popper claim, all such reasoning is unjustified. The first option is, of course, also taken by certain religious zealots who claim that science is worthless, that views about the nature of the world should be taken on faith in religious texts and doctrines.

I will assume, without argument, that Newton's aim of establishing a scientific law to explain a range of phenomena is legitimate and can, in certain situations, be achieved. Can it be achieved by following Newton's strategy? Unless you can show that causal and inductive reasoning of the sort involved in his rules can never be used to establish any causal law, I will assume that Newton's strategy will work, depending on the empirical soundness of your arguments. You can argue from observed phenomena to causes, you can determine whether these causes exist, and are the same or different ones, you can determine whether the causes postulated really do cause the phenomena, and you can generalize your finding. You can do so legitimately, depending on how you argue.

The second way of attacking Newton would be taken by anyone who thinks that there are better strategies for establishing causal laws than following Newton's rules. Are there better strategies? Newton obviously thought not. He rejected, quite legitimately I think, two leading competitors of his day, Cartesian rationalism and hypothetico-deductivisim. There are other legitimate scientific methods, as I will argue in the next two chapters, but some of these have quite different aims from Newton's—aims that are perfectly reasonable. There is also a legitimate extension of Newton's methodology proposed by Newton himself, and by Mill in the nineteenth century (Mill's "deductive method"), which I will discuss in chapter 3. And there are scientific methods that have been proposed with aims similar to Newton's but with rules very different from his. I will discuss one such, "inference to the best explanation," in the chapter that follows, arguing that if you follow the strategy it proposes, and do so correctly, you will not thereby establish a hypothesis or be justified in believing you have.

The third type of criticism mentioned above will no doubt produce sympathy in those who point out that no respectable scientist other than Newton ever used his rules—not because it is impossible to do so, but because they are empty and trivial (as the chess example above is supposed to show). They don't really tell scientists what to do in any important way. For example, the third rule tells you to generalize when you are warranted, but doesn't tell you when you are warranted or how to generalize.

In response, it should be emphasized that Newton's rules are part of a general strategy for trying to establish a scientific law to explain a set of phenomena. If the strategy and accompanying rules were as trivial as the chess rule to checkmate your opponent when you can, then there would seem to be no point in Newton's rejecting alternative strategies such as Cartesian a priorism or hypothetico-deductivism. Presumably, all chess strategies whose aim is winning would include this (trivial) chess rule. Presumably all reasonable scientific strategies aiming to establish a law would include the rule "generalize when you can." But that is not a fair representation of Newton's third rule, or of his general strategy. It is by no means trivial and uninformative to say, as Newton does, that one should try to establish the law of gravity by starting with observed phenomena, empirically inferring a common cause of these phenomena, and then attempting to inductively generalize as broadly as possible to a law that governs this cause and applies to the phenomena with which he began and others as well. This is part of his inductive strategy, which is very different from strategies advocated by two of his most famous methodological opponents, Descartes and Huygens. In trying to establish his laws of motion, Descartes begins with intuitions about God, from which he deduces his laws of motion. Huygens, using a version of the hypothetico-deductive method, begins with unproved assumptions about the wave nature of light, from which he deduces known properties of light. Newton's strategy precludes these.

In the present chapter I have provided a way of interpreting Newton's strategy and rules that captures much of what he says and does in deriving and defending the law of gravity, that makes the strategy and rules reasonable, and that avoids standard criticisms of his methodology. I will begin the next chapter with a discussion of extensions of Newton's methodological strategy.

Chapter 3

Newtonian Extensions, a Rival, Justifying Induction, and Evidence

In the present chapter I will examine extensions of Newton's methods; a powerful rival—inference to the best explanation; the Humean question of how to justify causal and inductive reasoning generally; and whether the problem of evidence introduced in chapter 1 can be solved by applying rules such as Newton's to particular evidential claims.

1. Extensions of Newton's Method

At the end of his book *Opticks*, Newton distinguishes two methods, which he calls "analysis" and "synthesis" (or "composition"):

> As in Mathematicks, so in Natural Philosophy, the Investigation of difficult Things by the Method of Analysis, ought ever to precede the Method of Composition. The Analysis consists in making Experiments and Observations, and in drawing general Conclusions from them by Induction, and admitting of no Objections against the Conclusions, but such as are taken from Experiments, or other certain Truths. For Hypotheses are not to be regarded in experimental philosophy. And although the arguing from Experiments and Observations by Induction be

no Demonstration of general Conclusions; yet it is the best way of arguing which the Nature of Things admits of, and may be looked upon as so much the stronger, by how much the Induction is more general.... By this way of Analysis we may proceed from Compounds to Ingredients, from Effects to their Causes, and from particular Causes to more general ones, till the Argument end in the most general. This is the Method of Analysis: And the Synthesis consists in assuming the Causes discover'd, and establish'd as Principles, and by them explaining the Phenomena proceeding from them, and proving the Explanations.[1]

Newton's method of analysis is, in effect, the method or strategy I have presented in chapter 2, one governed by the four methodological rules for causal and inductive reasoning from phenomena. The method of synthesis adds the idea that you take the general causal law you have arrived at by the method of analysis, together with other propositions already established by this method, and use these to explain phenomena and "prove" the explanation. By "prove" I will assume Newton means "test," rather than "establish." In the *Principia*, the "synthesis" is carried out in a series of "Problems" in Book 3. For example, Problem 6 is "To find the forces of the sun that perturb the motions of the moon." Newton offers a number of computations, following which he writes:

> I wished to show by these computations of the lunar motions that the lunar motions can be computed from their causes by the theory of gravity.[2]

---

1. Isaac Newton, *Opticks* (New York: Dover, 1970), pp. 404–5.
2. Isaac Newton, *The Principia*, trans. I. Bernard Cohen and Anne Whitman (Berkeley: University of California Press, 1999), p. 869.

EVIDENCE AND METHOD

These motions are thereby explained using the law of gravity, which is accomplished by applying the law, together with other established propositions, not only to phenomena with which Newton began (including, e.g., lunar motion) but also to others that were not used in generating it (e.g., the tides), and seeing whether such phenomena in fact obtain in the manner the law prescribes.[3] This can result in the sort of refinement procedure I noted in chapter 2. As both Smith and Harper[4] emphasize, it may enable you to formulate more precisely the phenomena (e.g., deviations from Kepler's laws applied to motions of planets and moons) from which the law was derived, or at least to show that such a refinement is necessary if you seek "exact truth."

How does one "prove" or "test" the explanation? Presumably by determining whether the computation from the law, plus the other established propositions, will indeed yield an answer to the question posed about the phenomena in question, and if the phenomena are observed to be "very nearly" as computed. If this turns out to be the case, what can be concluded? Newton does not explicitly say. To be Popperian about it,[5] we might conclude at least that the law has not been refuted by these computations. Using a probability idea, we might also conclude that the law, established with high probability on the basis of the initial phenomena, is now shown to have at least as high a probability on the basis of a broader range of phenomena than before.[6] We might also say that if the law can be used

---

3. As noted in chapter 2, section 3, Newton mentions this idea in the passage quoted from his Preface to the First Edition of the *Principia*, although he does not there call it "synthesis."
4. George E. Smith, "The Methodology of the *Principia*," and William Harper, "Newton's Argument for Universal Gravitation," in I. Bernard Cohen and George E. Smith, eds., *The Cambridge Companion to Newton* (Cambridge, UK: Cambridge University Press, 2002).
5. Karl Popper, *The Logic of Scientific Discovery* (New York: Basic Books, 1959), chapter 1.
6. This is based on the probability theorem that if $p(h/e) = k$, and if h entails $e_1, \ldots, e_n$, then $p(h/e \,\&\, e_1, \ldots, e_n)$ is greater than or equal to k.

to explain a broader range of phenomena than those we started with, this makes it a more powerful, more unifying law, even if it doesn't raise its probability. Finally, if we can make causal-inductive inferences from these new phenomena to the law—that is, empirically defensible ones satisfying Newton's rules—then, by Newton's standards, that should serve to increase the believability or probability of the law.

In the mid-nineteenth century, John Stuart Mill developed a method that in effect combines Newtonian analysis and synthesis. He calls it the "deductive method." It applies to cases typical in the sciences when several different causes are responsible for a phenomenon. Mill writes:

> in order to discover the cause of any phenomenon by the Deductive Method, the process must consist of three parts—induction, ratiocination, and verification. Induction, (the place of which, however, may be supplied by a prior deduction,) to ascertain the laws of the causes; ratiocination, to compute from those laws how the causes will operate in the particular combination known to exist in the case in hand; verification, by comparing this calculated effect with the actual phenomena.[7]

Mill suggests that this is what Newton is in fact doing when he "proves the identity of gravity with the central force of the solar system." Mill's deductive method uses a basic idea expressed in Newton's four rules: we establish causal laws by causal-inductive reasoning (although both Newton and Mill allow that we may deductively derive a causal law from other laws arrived at inductively).

---

7. John Stuart Mill, *A System of Logic: Rationcinative and Inductive*, 8th ed. (London: Longmans, 1872), pp. 322–23.

What Mill stresses is the idea that the phenomena that we are concerned with may have several causes that are subject to different laws. If so, these laws will need to be combined in a process that involves calculation ("ratiocination") in order to derive phenomena whose existence we can test by experiment and observation. This testing is the stage of verification. The steps of ratiocination and verification are perhaps what Newton has in mind by the "method of synthesis."

When he speaks of "induction" in his first step, Mill means an inference he describes as follows:

> Induction, then, is that operation of the mind by which we infer that what we know to be true in a particular case or cases, will be true in all cases which resemble the former in certain assignable respects. In other words, Induction is the process by which we conclude that what is true of certain individuals of a class is true of the whole class, or that what is true at certain times will be true in similar circumstances at all times.[8]

Mill's idea here is more general than what Newton expresses in his inductive Rule 3, since the latter is restricted to bodies (and to qualities of those bodies that cannot be intended and remitted). And at least the second of Mill's definitions above speaks about generalizing from some to all members of a class without introducing any requirement that the sample class must have been observed or that it must include all observables, although it is clear from Mill's examples that he has in mind (at least) inferences from what has been observed.

---

8. Mill, *System of Logic*, p. 188.

NEWTONIAN EXTENSIONS

The important similarity between Mill's ideas and Newton's (as I am interpreting the latter) is that inductions involve generalizations, whether or not these are warranted. Mill explicitly allows that some inductions are bad ones because the sample is biased, or of insufficient size, or because other information impugns the induction. He explicitly rejects the idea of induction by simple enumeration, which Bacon called puerile, according to which you are warranted in making an induction simply because it is an inference from properties of the sample to the claim that these are also properties of the entire class. For Mill, whether an induction is warranted depends on empirical facts about the property involved and the sampling procedure. Essentially, that is the idea behind my formulation of Newton's strategy and rule 3. First, define induction the way Mill has done, recognizing that some inductions are good, some bad. Then, as Mill explicitly does, and as I am interpreting Newton, say that in attempting to arrive at a general causal law, if you have established that this law holds for all observed bodies (or objects, more generally), try to establish empirically that it holds for all bodies (objects).

Accordingly, an extended Newtonian method would involve adding a new rule to the other four, which, following Newton, I will call "synthesis." This is for cases in which you make causal-inductive inferences to several laws, which you seek to combine to explain certain phenomena:

(rule 5) *Synthesis*: Using rules 1–4, suppose you have determined (i) that phenomena in a given set have multiple causes, (ii) what these causes are, and (iii) what laws govern these causes. Then, assuming these laws can be combined, do so, and by calculation determine whether propositions describing the phenomena with which you began, as well as others, can be derived from this set. Determine by observation and experiment whether, and to what extent, actual phenomena that occur conform to the observational conclusions derived.

There is one additional extension of this methodology that I will mention: causal-inductive cases involving eliminative reasoning. Sometimes we are dealing with a type of phenomenon P that is not produced jointly by a number of different causes acting simultaneously, but that can be produced by any one of a number of different, mutually exclusive causes. In such cases we have a phenomenon of a certain type P that, using causal reasoning, we have inferred is sometimes caused by C_1, sometimes by C_2, and so on, and by no other causes we have observed. We make an induction that all instances of P are produced by one of these causes. Now we have observed a particular phenomenon of type P, and we show that it cannot be produced by any of these causes except for one, since the assumption that it is produced by any of the other causes leads to consequences incompatible with observations. So, given what we have observed, it must be the remaining cause.

A weak version of this strategy is employed by Newton himself in his *Opticks*, Book 3, when he presents an argument in favor of his particle theory of light. The argument considers two possible causes of known optical phenomena—particles and waves—and then points to optical phenomena such as diffraction that seem to be incompatible with the wave theory while being compatible with the particle theory. In the case of other known wave motions, such as sound and water waves, diffraction (or bending) into the shadow of an obstacle is observed, but no such diffraction was observed by Newton or others in the case of light. Newton's argument gave some reason for preferring one theory over the other, although it was not, and was not intended to be, conclusive in a manner that accords with the use of Newton's rules.[9] However, a century and a half later, in the early and mid-nineteenth century, wave theorists presented arguments in favor of the opposing wave theory that were of this causal-eliminative type and whose conclusions were considered established with the certainty required by Newton's methodology. The claim was that the only known causes of motion are via (classical) waves and particles, from which it was concluded that light is one or the other. But by this time, diffraction into the shadow had been observed, and wave theorists such as Fresnel proceeded to show how, in order to explain diffraction and certain other optical phenomena such as interference, the particle theory needed to introduce forces of a type never observed in nature, whereas the wave theory did not do so. From this it was concluded that the wave theory is probably true.[10]

In what follows, I will consider this type of reasoning (in its strong form illustrated by the nineteenth century wave theory) to be an eliminative version of Newtonian methodology. It employs causal and inductive reasoning (of the sort in Newton's rules 1–3) to the claim that the cause of the observed phenomena is a member of a certain set of

9. I understand Newtonian methodology to be one the use of which is intended to establish causal laws with "the highest evidence a proposition can have in this [experimental] philosophy" (Newton's expression). In chapter 4, I will classify Newton's argument for his particle theory of light as an example of Maxwell's "method of physical speculation," which does not have such certainty. In both cases, one can use causal-eliminative reasoning. Newton explicitly rejects eliminative reasoning of the following form: "hypotheses $h_1 \ldots h_n$ are possible, mutually exclusive, hypotheses to explain the phenomena; all but h_1 are refuted by observations; therefore h_1 is true." See I. B. Cohen, ed., *Isaac Newton's Papers and Letters on Natural Philosophy*, 2nd ed. (Cambridge, MA: Harvard University Press, 1978), p. 93. An argument of this form will not establish a hypothesis, since it contains no experimental justification for assuming that one or another of these hypotheses is true. The type of eliminative reasoning I am speaking of is different, since it does contain such experimental justification.
10. For a detailed probabilistic reconstruction of this argument, see my *Particles and Waves* (New York: Oxford University Press, 1991), chap. 3. Another example of this type of reasoning that I have discussed is Perrin's important argument for the existence of molecules from his experiments on Brownian motion. See my *The Book of Evidence* (New York: Oxford University Press, 2001), chap. 12, and my "Is there a Valid Experimental Argument for Scientific Realism," *Journal of Philosophy* 99 (2002): 470–95.

possible causes. And it employs an explanatory component (of a sort in the rule of synthesis 5) to show that all of the causes but one provide explanations of the phenomena that are either refuted or made very improbable, given those phenomena and others. If you use this type of reasoning in a way that is empirically defensible, given your epistemic situation, you will be justified in believing the conclusion.

I take rules 1–4, supplemented with rule 5, together with a causal-eliminative variant, to represent core features of Newtonian methodology. A useful way to examine this methodology is by comparing it with a powerful alternative that, in effect, claims that a law can and should be defended by using a method that fails to satisfy Newton's rules 1–4. To this I now turn.

2. Inference to the Best Explanation (IBE)

A chief rival to Newtonian methodology is a view that has become known as "inference to the best explanation."[11] The grandfather of such a view, and one of its greatest proponents, is William Whewell.[12] His position is something like hypothetico-deductivism, but more complex and sophisticated. It invokes two stages in attempting to establish a hypothesis: discovery and testing. In the first, "science begins with common observation of facts" (chap. 4, sec. 2), and proceeds from this to discover the idea of a hypothesis. To do so, Whewell writes, "there is introduced some general conception, which is given, not by the phenomena, but by the mind (chap. 5, sec. 2).

---

11. Gilbert Harman, who defended the view in the 1960s (he now rejects it), was the first to use this expression; "The Inference to the Best Explanation," *Philosophical Review* 64 (1965), pp. 529–33.
12. William Whewell, *The Philosophy of the Inductive Sciences* (New York: Johnson Reprint Corporation, from the 1847 ed.). Parts relevant to the discussion here are reprinted in my *Science Rules* (Baltimore: Johns Hopkins University Press, 2004), pp. 150–67.

This he calls a "colligation" of the facts. As an example, he cites the astronomer Kepler, who begins with observations (made by Tycho Brahe) of the positions of the planet Mars at various times of the year. What Kepler introduces, which is not given by the phenomena, is an idea that connects these observations—namely, that the Martian positions lie on an ellipse.[13] This Whewell calls an "act of invention," and he claims that it is a "conjecture." Its truth is not inferred from the facts it connects, since frequently the mind invents various conflicting hypotheses to colligate the same data before an inference is made to truth or falsity in the second, testing, stage.

In this second stage, Whewell imposes three conditions the satisfaction of which will justify an inference to the truth of the hypothesis. First, the hypothesis should, if true, explain phenomena that have been observed but also predict ones not yet observed.[14] Second, it should explain and predict phenomena "of a kind different from those which were contemplated in the formation of our hypothesis" (chap. 5, sec. 11). Whewell speaks of this as the "consilience of inductions." He cites Newton's law of gravity as an example because it explains all three of Kepler's laws, even though no connection between these laws had been contemplated before. Third, Whewell introduces a condition pertaining to the development of a scientific theory over time, a novel idea not discussed previously by scientists or philosophers. He notes that a hypothesis is usually part of a system, or theory, whose members are not framed all at once but are introduced, added to, and altered over time. Some theories become simpler, more unified, and more coherent in this process; these are the ones that turn out to be true. In

---

13. Whewell probably means that all the Martian positions, not just the observed ones, lie on an ellipse. Later, in chapter 4, section 2, when discussing Maxwell's method of physical analogy, I will understand colligation in a more restricted way to apply only to the observed points.

14. When Whewell and other IBE theorists use "explain," they don't mean "correctly explain." They require that the hypothesis, if true, (correctly) explain the phenomena.

false theories, the reverse is the case: they become more complex, less unified, less coherent. They introduce ad hoc hypotheses that do not fit in well with previous ones but are invoked simply to solve a particular isolated problem. I will call this Whewell's "coherence" requirement.

We can express these ideas as follows:

Whewell's Rule: You are justified in inferring that a hypothesis is true (or proved) if it explains and predicts not only phenomena of a type contemplated in its formation but also ones of types not so contemplated, and if it is part of a system that is coherent and over time has become even more so.

Whewell might be interpreted as a "holist" about inference, claiming that what you infer is not an individual hypothesis but, rather, an entire system of which the hypothesis is part. Or he might be interpreted as saying that you can infer an individual hypothesis, but only if it is part of an entire system that is coherent over time. I will construe him in the latter way. A major difference between his view and a simple version of hypothetico-deductivism is this: Whewell imposes the conditions of "consilience" and "coherence," while simple hypothetico-deductivism does not.

An important contemporary version of this view is expressed by Peter Lipton. He writes:

> Given our data and our background beliefs, we infer what would, if true, provide the best of the competing explanations we can generate of those data (so long as the best is good enough to make any inference at all). Far from explanation only coming on the scene after the inferential work is done, the core idea of Inference to the Best Explanation is that explanatory considerations are a guide to inference.[15]

---

15. Peter Lipton, *Inference to the Best Explanation*, 2nd ed. (London: Routledge, 2004), p. 56. Pages are given in text for selected quotations that follow.

Lipton's idea entails that "we have to produce a pool of explanations from which we infer the best one" (so long as it is "good enough" to infer), and what we infer is that it is true (p. 58). What counts as the "best"? Lipton distinguishes an explanation that is the "likeliest" or most probable, from one that is the "loveliest"—that is, one that "would, if correct, be the most explanatory or provide the most understanding" (p. 59). He opts for construing "best explanation" in terms of "loveliest" rather than "likeliest." Lipton does not spell out criteria of "loveliness," other than to say that "loveliness" pertains to the extent to which the explanation is explanatory and provides understanding. Since the term "lovely" is his, I will assume it involves the sorts of criteria one would use in determining how good an explanation is, and that these would include criteria Whewell has in mind in his second and third conditions for testing, viz., explanatory unification of a range of phenomena (consilience), simplicity, coherence, and so on. I will understand Lipton to be claiming that if, or to the extent to which, these are satisfied, the explanation will be not only "lovely" but "likely" as well.[16]

---

16. Although I will interpret Lipton's position in this way, he may have a somewhat looser notion in mind. He writes: "loveliness and likeliness will tend to go together, and indeed loveliness will be a guide to likeliness" (*Inference to the Best Explanation*, p. 61). In private correspondence, Marc Lange has suggested to me an example showing that loveliness cannot track likeliness in general. Newton's law of gravity restricted to the solar system would be less lovely than the unrestricted version, but more probable since the unrestricted law entails the restricted one but not conversely. Lipton could reply by saying that if the "restricted" law says that gravity applies only to the solar system, then the unrestricted law does not imply that. And if the restricted law says simply that all bodies in the solar system satisfy the law, then why is that less lovely than the more general law? (Is the claim that all planets obey Kepler's three laws less lovely than the claim that all planets and their satellites obey these laws?) In any case, Lipton's comparisons of loveliness are made between competing theories, not compatible ones. I will not pursue this example further, since I am unsure what Lipton would say here.

We can express this as follows:

Lipton's Rule (IBE): You are justified in inferring a hypothesis if the hypothesis, if true, would provide the loveliest explanation of the data of the available competing hypotheses that offer explanations of those data (assuming that the loveliest is good enough to make any inference at all).

The major difference between the methodologies of Whewell and Lipton, on the one hand, and Newton and Mill, on the other, is that Newton and Mill require that the hypothesis or law be inferred using inductive and causal reasoning of a sort described in Newton's four rules (the "method of analysis"), and in Mill's "Four Methods of Experimental Inquiry."[17] Whewell and Lipton do not. They move from observed phenomena directly to a hypothesis that potentially explains them. They consider such an inference justified if, or to the extent that, the hypothesis explains these phenomena and others, and if the explanations meet criteria of "loveliness." To be sure, the idea of explanation, as well as the verification of derived phenomena, is present in Newton's concept of "synthesis" and in Mill's concepts of "ratiocination" and "verification." But neither Newton nor Mill regards these as sufficient for inferring the truth of a hypothesis, even if one adds Whewellian or Liptonian "loveliness." As Mill explicitly claims in speaking about his "deductive method," Whewell omits the first step—the "inductive" one.[18] And, as Newton would no doubt say, "inference to the best explanation" omits the "method of analysis," leaving only the "method of synthesis" plus "loveliness." This is why Newton and Mill would reject

---

17. Mill, *System of Logic*, Book 3, chap. 8, p. 253ff.
18. Whewell does use the term "induction." But by this he means an inference directly from phenomena to an explanatory hypothesis that does not include reference to observed instances of that hypothesis.

"inference to the best explanation," as formulated by Whewell and Lipton. It is not enough.

Let us agree that all four of the methodologists would say that explanations (as well as predictions) of a variety of different phenomena, and subsequent verification of conclusions, are part of the scientific method they espouse. In addition, Newton and Mill require causal-inductive steps (following rules 1–4) to make it possible to establish a causal law. Whewell and Lipton omit such a requirement,[19] adding instead requirements some call "aesthetic" and Lipton calls "loveliness-making" (explanatory depth, unification, simplicity, consilience, coherence). It is these I focus on in what follows. Even if they make for a "lovely" or "lovelier" theory, do they make for a "likely" or "likelier" one? Are they epistemic criteria? Do they make for a theory we are justified in believing to be true?

3. Lipton's IBE Examined

In this section, I will focus on Lipton's position. It is, I believe, the best contemporary representative of this point of view, but it does need clarification. In the next section, I will discuss both Whewell and Lipton in broader terms. (Readers who want to approach the deeper philosophical issues about IBE more quickly and avoid particular problems about Lipton's formulation may wish to turn to the next section.)

There are several potential ambiguities in Lipton's position. His IBE rule as expressed above has an important conditional in it, viz., "if true." It states that I may infer a hypothesis when, *if true*, it

---

19. Indeed, Whewell explicitly rejects all four of Newton's rules, unless they can be construed to be saying what he (Whewell) wants to say about scientific method.

would provide the loveliest explanation of the data.[20] How should we understand this? Is the idea that a hypothesis requires truth in order to be sufficiently "lovely" to infer—that is, that truth is a necessary condition for loveliness? If so, you could have a hypothesis that unifies, simplifies, and so on (I will call this "potentially lovely")—indeed, more so than any of its competitors—but is not (actually) lovely, or sufficiently lovely to infer, since it is false. If so, this would make IBE inferences question-begging ones of the form

> (1) h is potentially the loveliest of competing explanations of the phenomena
> (2) h is true
> Therefore, h is the loveliest of the competitors (since (1) and (2) are the case)
> Therefore, h is true.

Obviously this won't do. What I take Lipton to be saying is that if a hypothesis is potentially the loveliest of competitors, then we may infer that it is true. In the argument above we can go directly from premise (1) to the second conclusion without requiring premise (2). In what follows, then, I will understand "lovely" to mean what I have called "potentially lovely," without a truth condition. If a hypothesis is the loveliest of the competitors (in this sense), then it is inferable, on Lipton's view. And we can understand its being the loveliest without requiring it to be true in order to be the loveliest.

Now let's consider Lipton's proviso "assuming that the loveliest is good enough to make any inference at all," stated at the end

---

20. Cf. n. 14.

of the rule in the previous section. His view might be construed as saying that if you are in a situation where you have several competing explanations, each of which is sufficiently probable to make an inference to it from the data, then you should infer the "loveliest" of these.[21] What counts as "sufficiently probable to make an inference"?

Suppose we say that an explanation should not be regarded as inferable if its probability is ½ or less, no matter how "lovely" it is. So if E is the "loveliest" of the competing explanations, and if E is inferable, then E's probability must be greater than ½. By the rules of probability this means that each of its competitors, however "lovely," must have a probability less than ½. So none of its competitors is even minimally inferable. That doesn't give Lipton what he wants.

Let's weaken the probabilistic standard for minimal inferability by requiring only that the probability of a minimally inferable explanation be greater than 0. Then we could have a situation in which the most beautiful explanation in the set of minimally inferable ones has an extremely low probability, even if it is the loveliest in the set. This seems unacceptable. Shall we be allowed to infer the truth of an explanation with extremely low probability just because it wins a beauty contest?

We need to replace the probabilistic versions of "minimally inferable" given above by some nonprobabilistic one. The obvious suggestion is "loveliness" itself. The idea is (1) that there is some threshold level of "loveliness" beyond which an inference to the truth of an explanation would be okay, but for the presence of the competitors in the set (call this "potential inferability"); and (2)

---

21. Lipton in fact mentions this possibility in *Inference to the Best Explanation*, p. 61.

that we are to infer the loveliest one in this set (call this "actual inferability"). I will suppose that this is Lipton's view. Some explanations are just too ugly to be considered as real competitors (they are not even potentially inferable). Others, although lovely enough to be potentially inferable competitors, are just not the loveliest, and so are not actually inferable.

Now the question becomes: Why should we accept the idea that "loveliness," of some degree, makes an explanation at least potentially, if not actually, inferable? There are two possibilities here. First, we might claim, as Lipton does, that there is a relationship between "loveliness" and "likeliness." In general, the higher the degree of the former, the higher the degree of the latter. Second, we might claim, as Lipton does not, that there is no such relationship, and it doesn't really matter. There are situations in which we are justified in inferring the truth of the loveliest explanation, even if its probability is very low, perhaps even lower than that of uglier explanations in the competing set.

The second possibility is not Lipton's choice, nor should it be. If the probability of an explanation is very low, that should be sufficient reason not to infer that it is true or to regard it as even potentially inferable. This is especially so if probability is construed, as I did in chapter 1, as objective epistemic probability—that is, as a measure of the degree of reasonableness of belief. It is not at all reasonable to believe an explanation to be true just on the grounds of its beauty unless beauty brings probability with it. As for the first possibility, Lipton offers no argument for the claim that likeliness is generally proportional to loveliness, except to say that it is, and that asking why is tantamount to asking Hume's question about why induction is justified (a question I will turn to in section 6).

Lipton's position, then, seems to be this: We have a set of known competing explanations, each of which is at least potentially inferable. We need to relate this to probability. Let's say that each explanation, if relativized to the same phenomena, has a probability greater than k, the threshold level for inferability. Let us suppose, however, that, if relativized not only to the phenomena but also to the additional fact that the competitors in the set have been proposed, all but one explanation has a probability less than k. One has a probability greater than k. That is the one that is actually inferable. If for k you are allowed to pick any number greater than zero, then we could have a situation in which the actually inferable (the loveliest) explanation has a probability that, although greater than k, is very low. That shouldn't be welcome to Lipton, for whom low probability should be a sufficient condition for noninference. So, we need to make k sufficiently high for inference. Say that inferability, whether actual or potential, requires a probability greater than one-half (more likely than not). Now we are in trouble, since the set of competitors is supposed to be potentially inferable. This means that relative to just the phenomena (but not to the fact that the other explanations are being considered), the probability of each is greater than one-half, which is impossible, since the theories are competitors.

We need two thresholds, one for "potential inferability" and one for "actual inferability." Let an explanation be potentially inferable only if its probability, given the phenomena, is greater than zero. And let it be actually inferable only if its probability, given both the phenomena and the fact that the competitors in question have been proposed, is greater than one-half. What Lipton has to assume is that the fact that an explanation is "lovelier" than any of its competitors in the set being considered boosts its probability from just being some number greater than zero, but not being greater than one-half, to being some number greater than one-half. How exactly does that work? Does the fact that other explanations in the comparison set are less beautiful than the most beautiful one make the latter even more beautiful, and hence more probable?

If Lipton assumes that, in general, probability is proportional to beauty, then to get what he wants he must assume that (a) the existence of a comparison class with less beauty can increase the beauty of the most beautiful in the class; and (b) that this will generally increase the probability of the most beautiful, to make it sufficiently high to infer. Perhaps citing a comparison class can make us realize that one explanation really is more beautiful than we thought before considering that comparison class. But can it really make it more beautiful in the objective sense that Lipton requires? So far as I can see, this is what he needs to make his case that the loveliest explanation in the required comparison set is sufficiently probable to infer. He does not show that this is the case, nor does it seem plausible to say that this is simply Hume's problem of justifying inductive inferences.

There are two additional clarifications needed in Lipton's position. First, suppose we have a set of phenomena P_1 for which explanation E_1 is the loveliest explanation in the set of competing explanations of P_1. So by IBE, we are to infer the truth of explanation E_1. Now there may be a different set of phenomena P_2 for which E_2 is the loveliest explanation in the set of competing explanations of P_2. So we are to infer E_2. But E_1 and E_2 may be incompatible, so we can't infer both.[22] To avoid such cases Lipton, like Harman

---

22. One among many scientific examples is Rutherford's explanation of his experimental results in the scattering of alpha particles by thin gold strips. The explanation involves postulating a small nucleus in the atom, surrounded by rotating electrons. This explanation may have been more beautiful than any competitors, but it was inconsistent with classical electrodynamics, which was more beautiful in explaining other electrical phenomena outside the atom.

before him,[23] and possibly like Whewell, needs to go "global" and adopt some version of holism. What we infer is not a single explanation of a set of phenomena, but an entire theoretical system that offers the best (loveliest) explanation for all the phenomena we are trying to explain.[24]

Finally, Lipton's rule IBE, as formulated in the previous section, is a rule indicating when one is justified in inferring a hypothesis. I will understand the concept of justification here to be relativized to some epistemic situation. A person in a certain epistemic situation ES is justified in inferring a hypothesis under conditions of the sort Lipton gives (whether in his original formulation or in my variants). One of these conditions will make reference to competing hypotheses available to those in ES.

4. IBE Examined More Generally

A fundamental question that must now be raised is this: Why suppose that the virtues of "loveliness" extolled by Lipton and Whewell—virtues such as explanatory depth, simplicity, consilience, and coherence—have epistemic value? Why suppose that they provide, or help to provide, a good reason to believe that a theory is true? Both

---

23. Gilbert Harman, *Thought* (Princeton, NJ: Princeton University Press, 1973).
24. Newton and Mill, by contrast, do not, and do not need to, espouse holism. To begin with, their inference rules, as I am interpreting them, are pragmatic, telling us what to try to do, not what to infer from what or what we are epistemically justified in inferring. Moreover, when a Newtonian or Millian inductive or causal inference is made, the inference is not a holistic one to an entire theoretical set of claims but, rather, to an individual causal law, although depending on the circumstances, the inference may be defended by appeal to empirical background facts not in the premises of the inference. See "The War on Induction," in my *Evidence, Explanation, and Realism* (New York: Oxford University Press, 2010), esp. pp. 80–83.

Lipton and Whewell assume that this is the case, but do not provide any reason for doing so.

I will assume, as I did at the end of the previous section, that epistemic value for IBE theorists is to be relativized to an epistemic situation. The question then becomes: Why suppose that the fact that a theory being considered by someone in a given epistemic situation ES satisfies criteria of "loveliness" constitutes a good reason for a person in that ES to believe that the theory is true? Or, to put this in a more Liptonian way, what arguments might be given to support the claim that "loveliness" is generally proportional to "likeliness" (understood as relativized to an ES)? Here are some possibilities.

1. *No Miracle Argument.* Suppose that a theory that explains various phenomena has the virtues the IBE theorist extols in more abundance than any known competitor. This calls for an explanation (it's no miracle). The best (loveliest) explanation of this fact is that the theory is true. Therefore, we can infer the truth of a theory or explanation from its having these virtues of "loveliness."[25]

Response. This begs the question, since we are using IBE to get from "this is the loveliest explanation of the loveliness of the theory" to "the explanation is true," and hence to "the theory is true."[26] We are saying that since this theory exhibits explanatory depth, simplicity, consilience, and coherence, the loveliest explanation of this fact is that the theory is true—where the inference to truth is justified by IBE.

---

25. Perhaps Whewell is thinking of such an argument when he introduces the idea of "consilience" and writes: "The instances in which this [consilience] has occurred, indeed, impress us with a conviction that the truth of our hypothesis is certain. No accident could give rise to such an extraordinary coincidence" (in Achinstein, *Science Rules*, pp. 160–61).
26. See Arthur Fine, *The Shakey Game* (Chicago: University of Chicago Press, 1986), p. 114.

2. *Ontological Argument.* Newton, we have already seen, makes the claim "Nature is simple" in defense of his Rules 1 and 3. Simplicity is one of the criteria of "loveliness," so let's extend the idea to say that "nature is lovely." Since nature is lovely, it might be said, a lovely theory has a better chance of being true than an ugly one, and the loveliest theory has the best chance.

Response. There are two problems with this. First, how are we supposed to know or find out that nature is lovely? If we use IBE, we beg the question once again. Second, even if nature is "lovely" (whatever that is supposed to mean about the world, by contrast to lovely theories), why suppose that a lovely theory, even the loveliest among competing theories, will get it right about the loveliness in nature? George Washington may have been very good looking, but why suppose that a lovely picture of him, even the loveliest, is in fact accurate, even if the artist intended it to be accurate? There are many different possible worlds that are lovely—possible worlds with different sorts of entities and laws governing them. But the fact that some theory is the loveliest of its competitors is no reason to suppose that it picks out the actual lovely world.[27]

3. *Inductive Argument.* Theories that are "lovely" (or lovelier than their competitors) have turned out to be true much more often than theories that are not lovely (or not lovelier than their competitors).

27. I would offer a similar response to a position taken by Thomas Nagel in his book *Mind and Cosmos* (New York: Oxford University Press, 2012). Nagel claims that "science is driven by the assumption that nature is intelligible. . . . So when we prefer one explanation of the same data to another because it is simpler and makes fewer arbitrary assumptions, that is not just an aesthetic preference; it is because we think the explanation that gives greater understanding is more likely to be true, just for that reason" (pp. 16–17). Even if science is "driven" by that assumption, how are we supposed to determine that the assumption is correct? And even if nature is "intelligible," why suppose that an intelligible (simple, coherent) explanation or theory, even the most intelligible one we have, gets it right about the intelligibility of nature?

So historically speaking, loveliness has a better chance of getting it right. Loveliness has a good track record.[28]

Response. As Larry Laudan has emphasized,[29] if you look at the success of theories historically, using any criterion of goodness, including "loveliness," the induction will be pessimistic. Indeed, even Whewell's two favorite theories (Newtonian gravitation and the wave theory of light) turned out to be false, despite the fact that, according to him, they completely satisfied the requirements of consilience and coherence.

4. *Habit.* We can't help but believe that a "lovely" theory (or one "lovelier than any known competitor") is true. It is just a habit of the mind. Get used to it.[30]

---

28. James McAllister, in his book *Beauty and Revolution in Science* (Ithaca, NY: Cornell University Press, 1996), has a chapter entitled "The Relation of Beauty to Truth," in which he discusses various scientists who have expressed the idea that beauty tracks truth. One such scientist is Roger Penrose, who seems to be taking the inductive stance when he writes "So often, in fact, it turns out that the more attractive possibility is the true one" (Roger Penrose, "The Role of Aesthetics in Pure and Applied Mathematical Research," *Bulletin of the Institute of Mathematics and its Applications* 10 (1974): 266–71, quote p. 267). Indeed, Whewell himself makes a claim about consilience that suggests this idea: "No example can be pointed out, in the whole history of science, so far as I am aware, in which this Consilience of Inductions has given testimony in favour of an hypothesis afterwards discovered to be false" (Achinstein, *Science Rules*, p. 162). Another scientist, also quoted by McAllister (p. 15), is P. A. M. Dirac, who proposes a similar idea when he writes "It seems that if one is working from the point of view of getting beauty in one's equations, and if one has a really sound insight, one is on a sure line of progress." Dirac, speaking of the general theory of relativity, stated that "it is the essential beauty of the theory which I feel is the real reason for believing in it" (as quoted in McAllister, p. 16).
29. Larry Laudan, "A Confutation of Convergent Realism," in Jarrett Leplin, ed., *Scientific Realism* (Berkeley: University of California Press, 1984), pp. 218–49.
30. McAllister, *Beauty and Revolution*, p. 90, quotes Heisenberg: "If nature leads us to mathematical forms of great simplicity and beauty—by forms I am referring to coherent systems of hypotheses, axioms, etc. . . . we cannot help thinking that they are 'true,' that they reveal a genuine feature of nature." This idea was suggested by Duhem, who claimed that the simpler and better organized the theory, the greater our "conviction" that it represents reality, even though science cannot prove that this conviction is true (Pierre Duhem, *The Aim and Structure of Physical Theory*, reprinted in Achinstein, *Science Rules*, p. 275).

Response. As in the case of Hume's appeal to the idea that induction is a habit of the mind, this is no justification of the practice of inferring truth from beauty. It is at best an acknowledgment of something that cannot be helped. But is it even that? As with induction, there are some inductions we make, others we don't. Some are good, some not. We are not in the habit of making just any induction whatever. Similarly, some "lovely" theories are inferred to be true, some not. And even when a lovely one is inferred, it doesn't follow that it is inferred to be true on grounds that it is lovely. Loveliness may be a virtue that we can't help but want without being an epistemic virtue.

5. *Semantic Argument.* A "lovely theory" (or one "lovelier theory than any known competitor") is likely to be true just in virtue of the meaning of the expression "lovely theory" (i.e., in virtue of the meanings of "simple," "consilient," "coherent," etc.).

Response: Thinking how this could be so, in another work[31] I constructed probabilistic definitions of Whewell's ideas of consilience and coherence to see whether a theory satisfying these definitions would thereby be highly probable, or at least more probable than one that failed to satisfy these definitions. My conclusion was that neither of these consequences obtains. Without such definitions, but using a probabilistic argument, Gregory Morgan argues for the same null-result.[32] Instead of pursuing such technical probabilistic definitions and arguments here, I will approach the issue with an example.

---

31. Achinstein, *Particles and Waves*, pp. 123–33.
32. Gregory J. Morgan, "Achinstein and Whewell on Theoretical Coherence," in Gregory J. Morgan, ed., *Philosophy of Science Matters: The Philosophy of Peter Achinstein* (New York: Oxford University Press, 2011), pp. 151–63.

In Part II, section 36, of his *Principles of Philosophy*, Descartes writes:

> After considering the nature of motion, we must treat of its cause; in fact, of two sorts of causes. First, the universal and primary cause—the general cause of all the motions in the universe; secondly the particular cause that makes any given piece of matter assume a motion that it had not before.
>
> As regards the general cause, it seems clear to me that it can be none other than God himself. He created matter along with motion and rest in the beginning, and now, merely by his ordinary co-operation, he preserves just the quantity of motion and rest in the material world that he put there in the beginning.... Further, we conceive it as belonging to God's perfection, not only that he should in himself be unchangeable, but also that his operation should occur in a supremely constant and unchangeable manner.... Consequently it is most reasonable to hold that, from the mere fact that God gave pieces of matter various movements at their first creation, and that he now preserves all this matter in being in the same way as he first created it, he must likewise always preserve in it the same quantity of motion.[33]

Descartes proceeds to argue that "from God's immutability" we can also know certain rules or natural laws that are the secondary, particular causes of the various motions we see in different bodies. He then claims that he can derive three laws of motion (two of which involve inertia, the third interactions between bodies).

Now my interest here is not in Descartes' arguments for God, for God's immutability, or for the three laws of motion. It is, rather,

---

33. Reprinted in Achinstein, *Science Rules*, pp. 40–41.

in the aesthetic character of the theory itself. The theory is pretty "lovely," I would say—at least with respect to a range of phenomena available to Descartes. It exhibits simplicity: one God, not many, to explain phenomena of motion. It unifies: different kinds of motions (rectilinear, circular, motions after collisions) are all explained by the hypotheses. It explains two of the most important laws in physics, the law of conservation of motion (or momentum, in Descartes' version of momentum) and the law of inertia. For Lipton this theory should be "lovely" because it provides a deep (simple, unifying) understanding of why the various laws mentioned obtain: they hold because God created and sustains matter and motion, and does so in the simplest way.

Suppose that, prior to Newton (and his rejection of Descartes' idea of momentum and his laws of collisions), this is the loveliest available explanation of various observed phenomena of motion. Would it be a semantic contradiction to say that even though the theory is lovely in the ways described above—lovelier than any competitor known at the time—it is not necessarily likely, or necessarily worthy of belief by those in Descartes' epistemic situation? It seems perfectly appropriate to remark: "Yes, I agree it is lovely in the ways indicated, but is it likely to be true?" This is not like saying: "Yes, I agree that this figure is the locus of all points equidistant from a given point, but is it really a circle?" This question is settled by an appeal to a definition. The former one is not, at least not by appeal to any definition of which I am aware or can produce.

Descartes, of course, does not make an inference to his theory using loveliness arguments of the sort above. His actual grounds for believing the theory to be true have to do with a priori reasons he gives for believing that God exists, that God created and sustains matter and motion, and that God conserves the total quantity of motion in the universe. For Descartes in the general case of motion, as well as for Newton in the case of planetary motions, if you introduce a cause to explain phenomena you need to supply reasons to believe that this cause exists and that it operates the way you say it does. The cause and laws governing it that you introduce are believable not in virtue of the fact that they are simple, unifying, and so on, but in virtue of the fact that you have presented arguments (empirical for Newton, a priori for Descartes) showing that the cause exists, and that it causes the phenomena in question and does so in accordance with the law(s) you introduce.

Another example of a lovely theory, which I will develop in the next chapter for a different purpose, is Maxwell's 1860 kinetic theory of gases. That theory explains a range of known phenomena by assuming that spherical molecules exist that exert only contact forces and that obey Newtonian dynamics. But in 1860, Maxwell offers no reasons to infer that these particular assumptions are true. Indeed, he regards them as arbitrary mechanical assumptions to be used for the purpose of seeing whether mechanical explanations of known gaseous phenomena are even possible. And he explicitly refuses to infer the truth of his theory, even though its explanations of various phenomena satisfy criteria of the sort IBE theorists cherish. If loveliness semantically entailed likeliness, then you would think that Maxwell would have inferred the truth of his theory from the fact that it explained and unified many gaseous phenomena.

Finally, I propose to raise a question about how best to understand the claim that a certain explanation is lovely or the loveliest one in a group. Is this an empirical or an a priori claim? And, whatever it is, is the inference from this claim to the conclusion that the explanation is true or probably true an a priori or an empirical inference? I suspect that the answer in both cases is supposed to be "a priori." For both Whewell and Lipton, as I understand them, whether and to what extent an explanation is "lovely" is completely an a priori matter, not an empirical one. Given the empirical phenomena explained, we simply examine the explanation to see how unifying, simple, and coherent it is with respect to those phenomena, in comparison with other competing explanations. No experiments or empirical observations are necessary to determine this. "Ratiocination" (to use Mill's term) is all that is required. And, if it is the loveliest explanation, then, I assume an IBE theorist will say that an inference to the claim that it is probably true requires no empirical justification. Let's put an IBE argument in this form:

Inference to the Best Explanation
Premise 1: Phenomena P, some of which were predicted by explanation (theory, hypothesis) T, have been observed.
Premise 2: T is the loveliest explanation of phenomena P in the set of considered competing explanations.
Therefore (probably),
Conclusion: T is true.

I take the IBE theorist to be saying that Premise 2 is a priori, as is the inference from the set of premises to the conclusion. To be sure, to establish Premise 2 as true may take a good deal of ratiocination. But it is all a priori. And, of course, whether the conclusion is true is an empirical matter. But whether the inference from the premises to the conclusion is justified is not.

Contrast this with inductive and causal inferences. Let's put these forms of inference as follows:

Inductive Generalization
Premise: All observed As are Bs.
Therefore (probably),
Conclusion: All As are Bs.

Causal Inference (Newton)
Premise: These natural effects are of the same kind.
Therefore (probably),
Conclusion: They have the same cause.

Like IBE, whether the conclusions of these forms are true is an empirical matter. But unlike IBE, these argument forms contain no premise that is a priori. And unlike IBE, whether the inferences are justified is an empirical matter, not an a priori one. IBE theorists want to understand a scientific inference from empirical phenomena to an empirical explanation as an argument one of whose premises makes a priori "aesthetic" claims about the character of the explanation in relation to the phenomena, and whose conclusion is inferable a priori. Newtonian methodologists want to insist that causal and inductive inferences from phenomena to explanatory theories are empirically justified and do not require any a priori aesthetic premises. They demand to know why IBE theorists think that beauty (and only beauty) leads to probable truth. Beauty is no doubt a virtue of a theory, just not an epistemic one.

5. Can IBE Be Salvaged? (Sort of)

One possibility would be to adopt an alternative Lipton rejects and say that IBE is an inference to the "likeliest" explanation, not the "loveliest." The idea would still be that "we infer what would, if true, provide the best of the competing explanations of those data" (Lipton), but now "the best" means "the most probable." In view of problems of the sort I noted in section 3, this would need to be reformulated in ways analogous to those I indicate in that section. But even assuming this can be done, there is a problem. To spell out this version we need to specify how we are to determine whether an explanation is the most probable. We can't use "loveliness" as our criterion for probability, for then we are back to where we started. If we use causal-inductive rules of the sort Newton and Mill have in mind, or if we use my pragmatic versions of these rules, then there is not much of a contrast between IBE and the methodologies of Newton and Mill once you add "synthesis" to Newton and Mill. An IBE theorist such as Whewell, who explicitly rejects Newton's rules and Mill's "deductive method," would object to this idea. So should Lipton, who places so much emphasis on "loveliness," since (as he recognizes) you can get probability without it.[34]

Let's keep "loveliness" and reformulate the basic ideas of "inference to the best explanation" in a manner analogous to the way I reformulated Newton's four rules, viz. as rules of strategy for achieving the aim of establishing a general causal law to explain a range of phenomena. We get something like this:

Strategy rule for IBE: If your aim is to establish a general causal law, try to show that the law offers the "loveliest" explanation of a range of phenomena (or try to show that, as a result of adding the law to a system you are using, you get the "loveliest" explanation for a range of phenomena).

Suppose you follow this strategy and you are warranted in your claim that the law or system in question is simpler, more unifying, and so on than any known competitor. That would show that your law or system is "lovelier" than others that are known, which would be a virtue of that law or system, but not an epistemic one. It would not give

---

34. "Sometimes the likeliest explanation is not very enlightening. It is extremely likely that smoking opium puts people to sleep because of its dormative powers…, but this is the very model of an unlovely explanation" (Lipton, *Inference to the Best Explanation*, p. 59).

you a reason for believing that the law or system is true or probable. By contrast, if you follow Newton's strategic rules as I formulated them, and are warranted in making the inferences you do, then you have a strong epistemic reason for believing the law or system. "Loveliness" makes for a better theory, not for a believable or more believable one.[35]

Is there a way to salvage "inference to the best explanation" and do so in a reasonable way? Yes, there is, and it's easy. Just change the aim in question from establishing a causal law that explains a range of phenomena to doing so in such a way as to satisfy some criterion of "loveliness" as well. What you want is a hypothesis or system that is both "likely" and "lovely." The idea might be put like this:

Combined Strategy: Follow the Newtonian rules 1–4 (as I formulated them in the previous chapter), as well as the rule of synthesis 5, doing so in such a way as to produce the "loveliest" system you can.

I regard this as a perfectly good strategy if your aim is to establish a law and system that are both "lovely" and "likely"—a perfectly good aim.[36] Many scientists, especially theoretical ones, frequently want beauty as well as truth. But these are different values. A true (or probable) theory may not be beautiful, and a beautiful one may not be true (or probable). Nor can we assume as a general rule that

---

35. Here I agree with Bas van Fraassen, *The Scientific Image* (Oxford, UK: Oxford University Press, 1980), p. 87; he writes, "When a theory is advocated, it is praised for many features other than empirical adequacy and strength: it is said to be mathematically elegant, simple, of great scope, complete in certain respects: *also* of wonderful use in unifying our account of hitherto disparate phenomena, and most of all, explanatory. . . . There are specifically human concerns, a function of our interests and pleasures, which make some theories more valuable or appealing to us than others. Values of this sort, however, provide reasons for using a theory, or contemplating it, whether or not we think it true, and cannot rationally guide our epistemic attitudes and decisions."
36. This is an aim a scientific realist (who wants truth) can have. Although he does not do so in the quote in the previous footnote, an anti-realist such as van Fraassen can express the point by saying that he wants a theory that is "empirically adequate" (one that "saves the phenomena") and one that is also "lovely," where these two values are unrelated.

if you have one you are more likely to have the other. However, if you follow the combined strategy above, and do so successfully, you will get both. Admittedly, this is not a strategy that would be advocated by supporters of IBE. They claim that producing the most beautiful explanation of various phenomena is sufficient for obtaining a causal law and a system incorporating such a law that is probably true. But they haven't shown that their strategy is likely to get truth; nor, I think, can they do so.

Earlier I spoke of Mill's "deductive method," which is composed of three parts: an induction to a set of laws, ratiocination or calculation of various consequences, and empirical verification of these consequences. The Combined Strategy, if applied in a manner that is empirically defensible, is a way of satisfying these three requirements. Indeed, if satisfied, it may well get you "loveliness" in addition, depending on the "loveliness" of the combined set of laws and of the derivations. Whether or not Mill would have regarded this as an extra benefit, what he rejected is the use of a method, which he attributed to Whewell, omitting the first, inductive step (or in my version, steps involving rules 1–4). We have no reason to regard a set of hypotheses, no matter how "lovely," as true or probable simply because these hypotheses have been shown to be the "loveliest" of all the known competitors.

Earlier I also claimed that if your aim is to establish a causal law (or set of them) to explain a range of phenomena, then, using Newton's rules, plus synthesis, in an empirically defensible way, you will achieve your aim. Using two opposing methods that Newton was aware of—Cartesian rationalism and hypothetico-deductivism—will not achieve your aim. Nor has it been shown by its defenders, nor I think can it be, that using "inference to the best explanation" in the versions I have considered will achieve your aim.

Can we conclude, therefore, that Newton wins the day? Just follow his rules, as I present them, and you will get what you want, provided that your aim is Newton's. Not so fast, you will say. The stumbling block here is following them "in an empirically defensible way." How do we do that, and how do we know we are justified in doing that in any given case? Don't we need a set of rules (not just strategic rules of the sort I have been urging) that tell us whether particular inductive or causal inferences are valid? And don't these rules require justification? This suggests a very old philosophical problem: the justification of induction, to which I will now turn.

6. The Justification of Inductive and Causal Inferences

The use of Newton's rules, as I present them, can lead to satisfying Newton's aim, if applied in an empirically defensible way. That is sufficient justification for employing them. But these are rules of strategy that tell you what you should try to do. They are not standard rules of inference that tell you what you can or should infer from what. The question, then, is this: When we have followed the rules, how do we know we have done so in an empirically defensible way? When Newton follows rules 1 and 2 and draws an inference from the observed motions of the planets and their satellites to the claim that the cause in each case is an inverse-square force, in fact the very same force (not many), how do we determine whether the inference is empirically defensible? I have already said that the inference is "pragmatically" justified, in virtue of being the sort of inference you need to satisfy the aim of arriving at a causal law to explain various phenomena. But why, if it is so, is it "epistemically" justified?

The usual answer is that if an inference is epistemically justified, it is so in virtue of satisfying some general rule(s) of inference. Indeed, this is the way Newton's rules are usually understood. But I have recommended a different way of viewing them, one that does not allow this answer. So, it might be claimed, given my view, we need a set of nonstrategic rules of inference in addition to the strategic ones I have outlined in order to determine whether the latter have been correctly applied. How are such rules to be justified? That is the usual problem of justifying inductive and causal reasoning.

The only justification that Newton gives for the rules, as he formulates them, is by an appeal to the claim that nature is simple (for his Rules 1 and 3). But what does this mean, and how does he know that it is true? Is the simplicity of nature a metaphysical principle not "deduced from the phenomena"? If so, Newton should reject it, based on his own methodological pronouncements. Is it an empirical principle? If so, it is one "deducible from the phenomena" using his rules. But that is circular. Is the only reasonable alternative to say with Hume that no inductive inferences are defensible, even though by habit we make them all the time? Or perhaps we should follow Karl Popper, who accepts Hume's skepticism but adds that if inductive reasoning is a habit it is a bad one that can be changed to allow only deductive reasoning.[37]

The mistake I see in the argument in the paragraph before the last one is to say that if an inference is empirically defensible, it must be so in virtue of some general rule of reasoning.[38] This is what I deny. Newton defends his inference to the claim that what causes the moon to stay in its orbit is the same cause as that which causes bodies

---

37. Popper, *Logic of Scientific Discovery*.
38. For a powerful critique of this idea, see John Norton, "A Little Survey of Induction," in Peter Achinstein, ed., *Scientific Evidence* (Baltimore: Johns Hopkins University Press, 2005), pp. 9–34; also Norton, "A Material Theory of Induction," *Philosophy of Science* 70 (2003): 647–70.

to fall to the earth, using an empirical argument that appeals to the fact that both the moon and falling bodies "fall" toward the earth, and do so with the same acceleration. To be sure, in accordance with rule 2, he then makes the inference to the claim that the causes of these motions are the same. But, I suggest, that claim is epistemically justified, or should be so regarded, not by rule 3 (either his, or my variant), but by what Newton has determined empirically about the bodies in question and their motions. It's not that rule 3 sanctions that particular inference. It depends on the empirical facts. Sometimes, depending on the facts, you can make an inference to a common cause, sometimes not. To be sure, the rule (my variant) tells you that in general you should try to make that sort of inference if you want to satisfy your aim. It pragmatically justifies an inference of that type. It does not epistemically justify that particular inference.

Suppose you agree that particular inductive and causal inferences are epistemically justified by the facts of the case, not by some general rules of inference. There are two fundamental questions you might ask. First, how are these justifying facts themselves epistemically justified? Second, what is the epistemic justification for making any inductive or causal inferences at all—that is, for the "practice" of induction? One who raises the first question might have in mind the following idea: either there are "basic" empirical claims that are self-evident and need no justification, and are such that all other claims can be justified by reference to these; or we get an infinite regress of justifications, and hence no justification.

These possibilities are not exhaustive, in my view. What needs justification is a contextual matter that depends on the epistemic situation of the justifier and that of the persons for whom the justification is given. Newton, for example, justifies his law of gravity by "deducing" it from his six Phenomena. The Phenomena are by no means self-evident claims that need no further justification. Indeed,

NEWTONIAN EXTENSIONS

Newton gives a justification for each, appealing to observational results established and accepted by members of the scientific community. He justifies what he needs to justify for the intended audience. That doesn't mean that the justification cannot be questioned. But to apply the lesson of rule 4 to this case, if there are questions about Newton's justification of his law of gravity, then by all means raise them. If new phenomena weaken or refute such a justification, then by all means question or reject the justification. But in the absence of this, if he has followed rules 1–4 in an empirically defensible manner, then, given that his epistemic situation is the one in question, he is justified in believing the conclusion he draws from the Phenomena he cites, without any infinite regress of justifications, without any appeal to self-evident empirical facts, and without assuming that the methodological rules provide an epistemic justification for this conclusion.

The second fundamental question above asks what epistemic justification there is for making any causal or inductive inferences at all—that is, for the practice of induction. For Hume, there is none. It is simply a habit we have. Yes, it is a habit we have, but it doesn't follow from this that no particular inductions are justified. Nor, because it is a habit, does it follow that our habit is to make any old inference from "all observed As are Bs" to "all As are Bs," or from "these effects are the same" to "their causes are the same." Our habit is much more sophisticated, and the particular inferences of these forms that we do make will depend on the As, Bs, causes, and effects in question. The "practice" of making inductive and causal inferences is a habit; it is what we do. As such, perhaps the only justification appropriate is "our survival," which is not an epistemic justification, even for the habit, but a pragmatic one. Any particular instance of that habit, any particular causal or inductive inference, is to be epistemically justified by the facts of the case.

7. Evidence

At the end of chapter 1 on evidence, I raised the following problem: Even if we accept my definitions of evidence as correct or plausible, how are we to apply such definitions to actual cases to determine whether e is evidence that h in such cases? This question is especially important for my definitions, since evidential claims, on my view, are, in general, not a priori but empirical.[39] If they were a priori, then in a given case whether e is evidence that h could be settled by "calculation." But if they are empirical, how are they to be settled? The question that I proposed to consider is whether particular evidential claims can be determined to be true by applying rules of a scientific method, such as those of Newton. It is not my claim that if methodological rules can be so employed, then Newton's rules are the only or the best ones suitable for the purpose. But since I have been discussing and interpreting these rules at some considerable length, and since they represent a well-known important empirical methodology, let us see whether they can be used for this purpose.

I will understand Newton's rules in the pragmatic way suggested. And I will assume that Newton has followed these rules and has arrived at his law of gravity from his six Phenomena in a way that is empirically defensible. Is it the case that the Phenomena in question constitute evidence, in one of my senses, that his law is true? My answer is yes, but this needs scrutiny because this answer is not as simple or as rosy as readers, especially my critics, may want.

---

39. Perhaps we can make some of them a priori by including "idealizations," but as suggested in chapter 1, that's not so easy. For example, we might say that the fact that John owns 95 percent of the tickets in an ideal lottery is evidence that he will win, where we understand an ideal lottery to be one in which one ticket will be drawn at random, no tickets will be destroyed, no owners will die, no ticket will be declared ineligible, nor will the entire lottery be declared illegal, etc. (Do we need to keep going?)

We need to ask what kind or kinds of evidence we will obtain, which depends on what "empirical defensibility" requires. There are various possibilities. In a strong sense, the possibility of which was noted in chapter 2, it requires something that is not relativized to a scientist or an epistemic situation. In this sense, if the argument from phenomena to a causal law is empirically defensible, it is so irrespective of anyone's beliefs or knowledge. I will also assume that if an argument is empirically defensible, its conclusion is highly probable (in a sense of "probable" not relativized to a scientist or epistemic situation), or at least more probable than not. Suppose, then, that using Newton's rules as I interpret them, we produce an empirically defensible argument in this sense from established phenomena e to a causal law h. Then the following conditions are satisfied:

1. Given the phenomena e, the probability is high (greater than $\frac{1}{2}$) that there is an explanatory connection between h and e. This is because, if we follow Newton's strategy, we will generate a causal law that will explain the phenomena, and we will do so in an empirically defensible way (in a nonrelativized sense). Hence, the probability will be high that there is an explanatory connection between the law and the phenomena.
2. e is true. Newton's strategy starts with phenomena that have been established as true. (Here I will follow Newton and consider as true propositions ones that are "exactly or very nearly true."[40])

---

40. Reported facts that are "very nearly true" can be evidence for hypotheses. The fact that Don Giovanni, in Mozart's opera by that name, seduced 1,000 Spanish women is ample evidence that he was a scoundrel, despite the fact that, as his servant Leporello points out in the Catalogue song, the precise number was 1,003. The fact that the orbits of the planets obey Kepler's second and third laws is evidence that they are subject to an inverse-square centripetal force directed toward the sun, despite the fact that the reported fact is not precisely, but very nearly, true.

3. *e* does not entail *h*. Otherwise, the argument leading from *e* to *h* will be an a priori one, not an empirical one subject to Newton's four rules.

If these conditions are satisfied, then, by my definitions in chapter 1, *e* will be (at least) potential evidence that *h*. If we add that *h* is true, and that there is an explanatory connection between *e* and *h*, then *e* will also be veridical evidence that *h* (in the strongest sense of veridical evidence that I introduced).

Consider now relativized senses of "empirically defensible"—first one, used in chapter 2, that is relativized to an epistemic situation ES. Suppose that, using Newton's rules, an argument to a causal law *h* is produced from the phenomena *e* that is empirically defensible, given the epistemic situation of those producing the argument (e.g., Newton's ES). Anyone in ES would be justified in believing the conclusion *h*, given the phenomena *e*. Then the conditions for what I have called ES-evidence are satisfied. That is, *e* is true, and anyone in ES is justified in believing that *e* is (probably) veridical evidence that *h*.

Finally, we might introduce a very weak relativized sense of "empirically defensible," one that entails only that the argument produced from phenomena to a causal law is or can be empirically defended by those who believe it establishes the law. Then the conditions for what I have called subjective evidence are satisfied: *e* is person X's subjective evidence that *h*, since X believes that *e* is (probably) veridical evidence that *h*, and X's reason for believing that *h* is true (or probable) is that *e* is true.

Accordingly, if we follow Newton's rules of strategy in an empirically defensible way, we will obtain evidence (in one or more of the senses I define) for the hypothesis we arrive at using that strategy. We will arrive at a causal law that will probably correctly explain the observed phenomena we start with (veridical or potential evidence).

Or we will arrive at a law that, given our epistemic situation, we are justified in believing will probably do this (ES-evidence). Or, at the very least, we will arrive at a law that we believe will probably do this (subjective evidence), whether we are justified or not. Again, however, it depends on whether we have followed that strategy in an empirically defensible way (in one of the relativized or unrelativized senses above). This is why I say that my answer to the problem of evidence raised in chapter 1 is not simple or rosy. Yes, following Newton's rules, we will obtain evidence, of one sort or another, but only if we follow the rules in an empirically defensible way, of one sort or another. One advantage of my definitions of evidence is that I can demonstrate the truth of the latter statement. But what we are lacking is a set of rules for determining whether we have followed Newton's rules in an empirically defensible way; and therefore, it might be said, we are still lacking a set of rules for determining how to apply my definitions of evidence to particular cases.

My response to these complaints is one I have given before. There are no rules of the sort being demanded. If you generalize in accordance with Newton's rule 3, whether you are empirically justified in doing so depends on facts about the sample, the sampling procedure, the population, the properties you are generalizing, and so forth. The (epistemic) defense of the generalization will not be by reference to a rule (such as induction by simple enumeration), but by appeal to facts of a type just noted. A similar point holds for causal inferences made in accordance with Newton's rules 1 and 2. Newton's rules, understood pragmatically, will yield a way to construct an argument from phenomena to a causal law, for which the phenomena may well be evidence in one or more of the senses I define. But you, or the scientist, will need to do empirical work to determine whether it is evidence. Rules of the sort philosophers, and some scientists, have sought will not tell you this.

Did Newton himself follow the rules, understood pragmatically, in an empirically defensible way? In the weakest relativized sense of "empirically defensible," he certainly did. He followed the rules, understood pragmatically, and gave empirical arguments that he believed established the law of gravity. He produced subjective evidence for the law. Given his epistemic situation, was he justified in believing the law? Did he have ES-evidence for his law? Newton was aware only of the motions of bodies that move with a speed much smaller than the speed of light; he did not know, nor was he in a position to know, that mass varies with velocity and that gravitational potential energies of the bodies he observed are small compared with mc^2—conditions not satisfied in regions close to large masses. He did not know about, nor was he in a position to know, the 38 arc seconds per century of Mercury's precession discovered by Le Verrier in 1859 that are inconsistent with his law. But given the observed phenomena he did have, he (his followers, and even contemporary historians of science) certainly believed that he was justified in believing the law, in a sense of "justified" required for ES-evidence. Since the law, construed in the most general sense intended by Newton, is in fact false, his evidence was not veridical evidence for that law. Nor was it potential evidence (which does not require the hypothesis to be true), since it is not the case that there probably is an explanatory connection between the law and the phenomena Newton cited. That is something we show empirically by reference to phenomena unavailable to Newton. In accordance with Newton's fourth rule, if he was justified in believing his law, given the phenomena in question (and more generally, given his epistemic situation), then he could justifiably infer the (approximate) truth of the law—until new phenomena are discovered that require amending or withdrawing the inference. That's what in fact happened, but long after Newton.

Portrait of Scottish physicist James Clerk Maxwell (1831–1879). Painting by an unknown artist. © Bettmann/CORBIS.

Chapter 4

What to Do If You Cannot Establish a Theory

Maxwell's Three Methods

1. Introduction

This book began with a problem about the objective empirical concepts of evidence I define: How can they be applied to actual scientific cases to determine whether some experimental results or observations count as evidence that a hypothesis or theory is true? The proposal was to consider whether a scientific method such as that embodied in Newton's rules could be used for this purpose. The answer is yes, but the issue is not an a priori one of the sort philosophers usually seek; it is empirical. Whether e is evidence that h, in one of the objective senses of evidence I propose, is a matter of scientific, not philosophical, inquiry.

Why do philosophers and methodologists of science focus so much attention on what is meant by evidence and what is required to produce it? One reason may have to do with what they regard as the aim of science. For example, Bas van Fraassen claims that defenders of both scientific realism and anti-realism (in senses of these terms he proposes) believe that science has an overarching aim.[1]

---

[1]. Bas van Fraassen, *The Scientific Image* (Oxford, UK: Oxford University Press, 1980), p. 8. Pages are given in text for selected quotations that follow.

The realist believes that "science aims to give us, in its theories, a literally true story of what the world is like; and acceptance of a scientific theory involves the belief that it is true" (p. 8). The anti-realist (or van Fraassen's "constructive empiricist") believes that "science aims to give us theories that are empirically adequate [that "save the phenomena"]; and acceptance of a theory involves as belief only that it is empirically adequate" (p. 12). If the aim of science is either truth or empirical adequacy, and if your aim as a scientist is to arrive at a theory you can accept, then obtaining evidence sufficient to convince you of the truth or empirical adequacy of your theory is what you want and need.

This focus on evidence may also derive from some influential views about scientific method. In chapter 2 of *Philosophy of Natural Science*, Carl G. Hempel embraces a version of hypothetico-deductivism, according to which scientific inquiry has two stages. First, there is the "invention" stage in which the scientist introduces a hypothesis to explain a set of observed phenomena. The hypothesis is a guess or conjecture that is not inferred from anything. The second stage, "testing," consists in drawing deductive inferences from the hypothesis to conclusions that are then tested by experiment and observation to see whether they are true or false. If the conclusions turn out to be true, the experimental results constitute evidence for the hypothesis, and depending on the number and variety of these positive results, they may establish that hypothesis. If one or more turn out to be false, the hypothesis must be modified or rejected. On such a view, these are the two stages that constitute scientific inquiry. There is no method or set of rules for inventing a theory in the first stage, while in the second there is a method—one utilizing deduction—but it is a method for confirming or establishing a theory by means of experimental evidence, or at least for

WHAT TO DO IF YOU CANNOT ESTABLISH A THEORY

testing it to show that it has not been experimentally refuted. A more elaborate version of this view, discussed in chapter 3, is due to William Whewell, who also distinguishes two stages of inquiry: one involves the formulation of a theory via "colligation," the other involves the application of criteria of "consilience" and "coherence" for testing and establishing the theory.

Are these the only types of "scientific inquiry" that philosophers of science concerned with scientific theorizing should think about? Suppose that you are not in a position to confirm or establish a theory (as true or empirically adequate). Either you have no theory at all to explain a set of established phenomena, or you do have a theory but you are not able to establish or even confirm it because, say, you don't know what experiments would do so, or even if you do, you cannot perform them. What, if anything, can you do as a scientist? One option, of course, is to do nothing. Don't speculate. This is Descartes' advice.[2] It is also Newton's advice at one point.[3] But there are some interesting alternatives that I will explore in this chapter. They are described by James Clerk Maxwell, probably the greatest theoretical physicist of the nineteenth century, and they occur in situations in which you don't have a theory that you can establish or even confirm

---

2. Descartes writes: "If in the series of things to be examined we come across something which our intellect is unable to intuit sufficiently well, we must stop at that point, and refrain from the superfluous task of examining the remaining items" (reprinted in Peter Achinstein, *Science Rules* [Baltimore: Johns Hopkins University Press, 2004], p. 25).
3. In a famous passage at the end of Book 3 of the *Principia* (Isaac Newton, *The Principia*, trans. I. Bernard Cohen and Anne Whitman [Berkeley: University of California Press, 1999]), Newton writes: "For whatever is not deduced from the phenomena must be called a hypothesis, and hypotheses, whether metaphysical or physical, or based on occult qualities, or mechanical, have no place in experimental philosophy." Interestingly, Newton doesn't always live by that philosophy. In the *Principia* itself, Newton introduces several unproved propositions that he calls "hypotheses." And in the *Opticks*, he introduces a set of "Queries" which are clearly hypotheses in his sense.

experimentally. Maxwell is concerned with several forms of scientific inquiry that are different from "discovering" hypotheses and "proving" them by means of evidence. These forms of inquiry, he thinks, like proving by evidence, are subject to methodological or strategic rules. Is he right? What are these methods? And are they philosophically and scientifically significant and plausible?

In this chapter I consider three methods introduced and employed by Maxwell. Taking them in the chronological order in which they were employed, the first he called a "method of physical analogy," the second "an exercise in mechanics," and the third a "method of physical speculation." As is the case with Newton, the three methods Maxwell introduces all involve "theorizing" with respect to a set of phenomena that have been established by observation. This "theorizing" is subject to three requirements that are the same for each method. First, a physical, rather than a merely mathematical, way of understanding the phenomena needs to be provided. Second, the "theorizing," although physical, should proceed in a precise way, using mathematics. Third, the "theorizing" should not be sketchy but worked out in detail. The physics that Maxwell actually produces using each of these three methods satisfies these requirements. And the methods he uses in doing so are different from any of those discussed in chapters 2 and 3.

2. A Method of Physical Analogy

In 1855, Maxwell published a paper entitled "On Faraday's Lines of Force."[4] It deals with the state of what he called "electrical science" in mid-nineteenth century, which consisted of unrelated

---

4. *The Scientific Papers of James Clerk Maxwell*, 2 vols., edited by W. D. Niven (New York: Dover, 1965), vol. 1, pp. 155–229. Pages cited in text for selected quotations are from vols. 1 and 2 as shown.

experimental laws concerning the distribution of electricity on the surface of conductors, the mutual attraction of conductors, and some phenomena involving magnets. But no theory was available that could unify and explain these various laws. Maxwell continues:

> The first process therefore in the effectual study of the science, must be one of simplification and reduction of the results of previous investigation to a form in which the mind can grasp them. The results of this simplification may take the form of a purely mathematical formula or of a physical hypothesis. In the first case we entirely lose sight of the phenomena to be explained; and though we may trace out the consequences of given laws, we can never obtain more extended views of the connexions of the subject. If, on the other hand, we adopt a physical hypothesis, we see the phenomena only through a medium, and are liable to that blindness to facts and rashness in assumption which a partial explanation encourages. We must therefore discover some method of investigation which allows the mind at every step to lay hold of a clear physical conception, without being committed to any theory founded on the physical science from which that conception is borrowed, so that it is neither drawn aside from the subject in pursuit of analytical subtleties, nor carried beyond the truth by a favourite hypothesis. In order to obtain physical ideas without adopting a physical theory we must make ourselves familiar with the existence of physical analogies. (1:155–6)

In electrical science in the 1850s, Maxwell finds a set of experimentally established laws that have not been related, simplified, or explained by some confirmed theory. In remedying this situation he rejects the use of purely mathematical formulas that might unify and simplify the laws but offer no physical basis for understanding them. And he rejects the use of physical hypotheses that can also unify and simplify the known laws but that are mere speculations that have not been experimentally established and so carry one beyond what can legitimately be inferred. What Maxwell proposes instead is the use of physical analogies, which he defines as follows:

> By a physical analogy I mean that partial similarity between the laws of one science and those of another which makes each of them illustrate the other. (1:156)

The best way to understand what Maxwell means here is to look at the actual physical analogy that he constructs.

Maxwell begins with Faraday's idea of "lines of force" to represent an electric or magnetic field. Suppose that a body is either electrically charged or magnetic. Then a small test body in its vicinity will be subjected to a force causing it to move in a certain direction, depending on the charges or the magnetic poles of the two bodies. Faraday represents this situation by means of lines of force that depict the direction of the force on the test particle at each point of space. Maxwell, in his physical analogy, seeks to extend this idea by providing a way to represent not just the direction of the force but its intensity or magnitude as well. He writes:

> If we consider these curves not as mere lines, but as fine tubes of variable section carrying an incompressible fluid, then, since the velocity of the fluid is inversely as the section of the

tube, we may make the velocity vary according to any given law, by regulating the section of the tube, and in this way we might represent the intensity of the force as well as its direction by the motion of the fluid in these tubes. This method of representing the intensity of a force by the velocity of an imaginary fluid in a tube is applicable to any conceivable system of forces, but it is capable of great simplification in the case in which the forces are such as can be explained by the hypothesis of attractions varying inversely as the square of the distance, such as those observed in electrical and magnetic phenomena. (1:158–59)

Maxwell constructs his analogy by describing a purely imaginary incompressible fluid flowing through tubes of varying section. The velocity of the fluid at a given point represents the electrical force at that point, and the direction of the tube represents the direction of the electric force. Particles of electricity are represented in the analogue as sources and sinks of fluid. And the electrical potential is represented by the pressure of the fluid. The velocity of the fluid at a distance r from a source will vary as $1/r^2$. We can tabulate these representations as follows:

Electric Field	Incompressible Fluid
Electrical force at a point in the field	Velocity of fluid at a point in the fluid
Particle of positive electricity	Source of fluid
Electrical potential at a point	Pressure of fluid at a point
Satisfies law that the force due to a charged particle at a distance r from the particle varies as $1/r^2$	Satisfies law that the velocity of fluid at a distance r from the source of fluid varies as $1/r^2$

The bulk of Maxwell's 75-page paper is spent working out this analogy mathematically, by showing how to derive equations governing the imaginary fluid that are analogues of ones governing electrical and magnetic fields. He sets his agenda as follows:

> I propose, then, first to describe a method by which the motion of such a fluid can be clearly conceived; secondly to trace the consequences of assuming certain conditions of motion, and to point out the application of the method to some of the less complicated phenomena of electricity, magnetism, and galvanism; and lastly to shew how by an extension of these methods, and the introduction of another idea due to Faraday, the laws of the attractions and inductive actions of magnets and currents may be clearly conceived, without making any assumptions as to the physical nature of electricity, or adding anything to that which has been already proved by experiment. (1:159)

Earlier I said that each of the methods Maxwell introduces is subject to three requirements: (1) the method must yield a way of physically understanding the phenomena; (2) the theorizing resulting from the use of the method must be formulated precisely, using mathematics; and (3) the theorizing must be worked out in detail. If one looks at his 1855 paper, it will be obvious that requirements 2 and 3 are satisfied. What about requirement 1, which, I think, is philosophically the most interesting? How can a physical analogy provide a physical way of understanding phenomena without explaining why they occur?

To answer, it will be useful to modify and extend some ideas of Whewell noted in chapter 3. Recall that Whewell (like Hempel) distinguishes two stages in establishing a hypothesis: discovery and testing. The former, which Whewell calls "colligation," involves representing the observed phenomena in a certain manner. Whewell's example is Kepler's representation of the observed positions of planet Mars as all lying on an ellipse. One way to interpret Whewell is as saying that the colligation applies the concept "lies on an elliptical orbit" to all points of the Martian orbit, observed as well as unobserved, so that colligation involves making an inductive inference. I propose to understand colligation in a more restricted way that does not commit one either to the inductive assumption that all the positions of Mars—unobserved as well as observed, future as well as past—lie on an ellipse, or to any theory about the cause of the observed phenomena. It is simply a representation of what has been observed, without any such implications.[5] This provides a way of understanding the phenomena—understanding *what* is occurring—by providing a way of seeing, conceptualizing, analyzing, or categorizing them that may be quite novel and fruitful, even if it doesn't explain *why* the phenomena occur or what causes them. Whewell stresses the importance of colligation, or classification in scientific theorizing, which he thinks is omitted or minimized in standard inductive and hypothetico-deductive accounts of scientific theorizing.[6]

---

5. Whether in fact Whewell's "colligations" do, or should, include both types, or perhaps only inductions, is controversial. There was a lengthy dispute on this issue between Whewell and Mill. Mill agrees with Whewell's view of colligation, but only if it is understood as involving no inductive or causal commitments. For Mill, it is simply a way of "describing" the observed data without inductive or causal reasoning. See my "The War on Induction," in Peter Achinstein, *Evidence, Explanation, and Realism* (New York: Oxford University Press, 2010), pp. 61–84.

6. He thinks of it as something that the mind imposes on the data. He stresses that the existence of this "mental act" is soon forgotten after the classification takes place, and we tend to think of classification in purely objective terms. I will not pursue this aspect of Whewell's account.

Kepler's way of understanding the various positions of Mars observed by Tycho Brahe as points on an ellipse, with the sun at one of the foci, may or may not be "correct" (the observed positions may not lie on an ellipse with the sun at one focus—indeed, the sun is not at a focus). It may not apply to unobserved phenomena of that type (to unobserved positions). Even if it does apply, it may not apply any more generally (e.g., to other planets). And it need not be a way that explains why the phenomena occur (representing the observed positions of Mars, or even the unobserved ones, as lying on an ellipse does not explain what causes those positions to be where they are). Whether a certain way of understanding the phenomena is "correct," or applies to unobserved phenomena of that type, or is generalizable to other phenomena of that type, or will lead to a correct explanation of the phenomena, are empirical matters to be settled by experiment and observation; a priori calculation will not suffice. According to Whewell, scientists frequently try different conceptualizations to determine their applicability to the observed phenomena. Nor, in general, are these conceptualizations simply "read off" from the data without calculation or inference. Kepler did not arrive at his conceptualization of the observed positions as lying on an ellipse around the sun simply by opening his eyes and looking. He had to calculate directions of sides on an earth-sun-Mars triangle from observations of the earth-Mars line.

One way to represent or conceptualize phenomena is the way Whewell is suggesting: provide a description of them that is to be taken more or less literally. But another way is to represent them by means of a description that is not to be taken literally at all, in this case an analogy. This is Maxwell's way. In an article on molecules in the *Encyclopedia Britannica*, Maxwell represents the

WHAT TO DO IF YOU CANNOT ESTABLISH A THEORY

motion of molecules in a gas by drawing an analogy between this and a swarm of bees. The bees in a swarm are the analogues of, and represent, a collection of molecules in a container of gas. And the random motions of the bees represent those of the molecules in the container. Maxwell, of course, is not supposing even for a moment that gases are composed of bees, but only that the molecules in gases behave like bees in certain respects. He introduces this analogy because, for those familiar with the motion of bees in a swarm, this will provide a way of understanding how molecules move in a container, even though it doesn't explain why this motion is the way it is.

Something of this sort appears in Maxwell's paper "On Faraday's Lines of Force." The electric field is represented by the incompressible fluid flowing through tubes of varying section. A particle of positive electricity is represented as a source of fluid, electrical potential at a point in the field by the pressure of the fluid at a point, and so forth. To those in Maxwell's day more familiar with hydrodynamics than with electrical theory, this is intended to provide a way of understanding electrical phenomena involving forces, currents, induction, and so forth, even though it doesn't explain what causes these phenomena.[7] However, there are several differences between this and the bee analogy. Bees exist, the incompressible fluid does not. For Maxwell this difference is unimportant. As long as we understand how the analogue is supposed to behave, it will serve its

---

7. Unlike the case with Kepler's colligation that all the observed points of the Martian orbit lie on an ellipse, which makes no inductive or causal commitments, here we are dealing with inductively established laws of electricity and magnetism. What Maxwell is doing is representing those laws, and the entities and properties involved, by describing an analogue system without assuming that such a system exists or that electrical or magnetic behavior is caused by any mechanism like that of the analogue system.

representational purpose, whether or not it exists. For Maxwell, one important difference between the present analogy and the bee analogy is that the former is expressed mathematically, the latter is not. This permits Maxwell to represent known quantitative laws governing the original system (the electric field) by quantitative laws governing the analogue (the fluid). So, for example, Coulomb's inverse-square law governing the force between electric charges is represented by a law relating the velocity of a fluid to its distance from the source. Indeed, for Maxwell, the idea of an analogy between the laws governing the original and those governing the analogue is so central that he incorporates it into his definition of a physical analogy (see above). Perhaps we can say that this is what is desired in a fully developed analogy.

This leads to a second important difference between Maxwell's bee analogy and his fluid analogy. The latter is very fully developed, the former is not. What advantages accrue from this? The electrical phenomena known to Maxwell were complex and varied, so if there is going to be a representation for all or most of them, it will have to be well developed. But I think that there are other reasons Maxwell had in mind. One is a mathematical solution of problems. In describing his analogue, Maxwell writes:

> It is not even a hypothetical fluid which is introduced to explain actual phenomena. It is merely a collection of imaginary properties which may be employed for establishing certain theorems in pure mathematics in a way more intelligible to many minds and more applicable to physical problems than that in which algebraic symbols alone are used. (1:160)

The idea seems to be this. We begin with a set of established laws in electricity, which we seek to organize and relate mathematically.

One way to do so is by employing an analogy in which the analogue is subject to laws of the same mathematical form. It may be simpler and more convenient to work out a mathematical problem first in that system, and then apply the solution to the original system.[8] Solving a mathematical problem in a system with which we are familiar may make it easier to solve the problem in the less familiar system. In the physical case of the sort of interest to Maxwell, it gives the mathematical problem a physical basis, which is an advantage to those who think in more concrete terms.

Another advantage Maxwell claims for a fully developed analogy is that it can be an important device in teaching and learning a science. He writes that it is

> not only convenient for teaching science in a pleasant and easy manner, but the recognition of the formal analogy between the two systems of ideas leads to a knowledge of both, more profound than could be obtained by studying each system separately. (2:219)

For these purposes the more fully worked out the analogy, the better the prospects for teaching and learning. Maxwell does recognize that there are

> some minds which can go on contemplating with satisfaction pure quantities presented to the eye by symbols, and to the mind in a form which none but mathematicians can conceive. (2:220)

---

8. Readers familiar with truth-functional logic who want to study elementary set theory can solve a problem in the latter by constructing the analogue in the former, and can then use a truth table to solve it.

But there are others that require that these quantities be represented physically with the use of physical illustrations and analogies. He continues:

> For the sake of persons of these different types, scientific truth should be presented in different forms, and should be regarded as equally scientific, whether it appears in the robust form and the vivid colouring of a physical illustration [or analogy] or in the tenuity and paleness of a symbolical expression. (2:220)

A final advantage of a fully developed analogy, one that Maxwell does not discuss, is that it may be useful in analogical reasoning. It may suggest the existence of as yet undiscovered laws and properties of the original system on the basis of ones present in the analogue system. It may even suggest underlying causes of the phenomena in the original system from ones governing the analogue system. This might be merely a suggestion—try it out, see whether such causes exist—rather than an inference to a probable cause. Or, depending on how similar the analogue system is to the original, and depending on whether the analogue system is real and known to exist and have the properties attributed to it, it may justify inferences of this sort. Perhaps Maxwell does not discuss the inferential value of analogies because, in his paper "On Faraday's Lines of Force," he explicitly wants to avoid inferring any hypotheses about the causes of electricity and magnetism from properties of his imaginary fluid. Since the fluid is imaginary, and subject to arbitrary assumptions, no such inferences from causal features of one to those of the other will be justified.

Whatever the case, if your aim, like Maxwell's, is to provide a way of understanding known phenomena without constructing a theory to explain what causes them to occur, and to do so in a way that enables known laws governing the phenomena to be organized and mathematically related so that problems can be solved mathematically, and so that the laws can be taught and learned more easily, Maxwell's method of physical analogy is one method to use. Maxwell, I think justifiably, regards the use of this method as a legitimate part of theoretical scientific activity when no theory is available that can causally explain the phenomena. Whether a particular physical analogy will be successful in the respects noted is an empirical question that depends on whether the phenomena can be so represented (e.g., on whether the phenomena are subject to the laws in question, and whether these laws are really similar in form to those governing the analogue), and on whether such a representation will in fact promote mathematical solutions, understanding, teaching, and learning. In this respect, Maxwell's method of physical analogy is similar to Newton's method of analysis. Whether some particular "theorizing" that results from using the method is successful in achieving the aims of the method is an empirical, not an a priori, issue.

Indeed, Maxwell's method, like Newton's, is best construed as a method of strategy, although the aim of each strategy is different. Newton's aim is to establish a general law that invokes a cause to explain a range of established phenomena. Maxwell's aim is to understand a range of established phenomena without invoking a causal theory to explain them. Newton's rules, under my interpretation, tell you to try to empirically determine whether the causes you introduce exist and are different or the same, to try to generalize the claim that some property holds for all observed bodies to the claim that it holds for all bodies, and so forth. Maxwell's "rules" (he doesn't call them that, but I will) tell you to try to draw an analogy between the system you are studying and a different one, and to do so in a way that is worked out in some detail and is mathematically precise. Newton's rules do not say what particular causes to introduce for a given set of phenomena, or how to generalize inductively to a law. And the fact that you have followed the rules does not mean that you will be successful in achieving Newton's aim, or even that you will be justified in believing that you have achieved that aim. To obtain the latter, you must follow the rules in an empirically defensible way. To obtain the former, nature must cooperate. Similarly, Maxwell's rules do not say what analogy to introduce for a given set of phenomena. And the fact that you have produced an analogy that is worked out in detail and mathematically precise doesn't mean that you have achieved the Maxwellian aim of providing a noncausal physical understanding of a set of phenomena. The "understanding" produced, though worked out and precise, may be inadequate in ways noted above.

3. An "Exercise in Mechanics"

In 1860, five years after "On Faraday's Lines of Force," Maxwell published a ground-breaking work, "Illustrations of the Dynamical Theory of Gases," his first paper on the kinetic-molecular theory of gases. In it he assumes that gases are composed of spherical molecules that move with uniform velocity in straight lines, except when they strike the sides of a container; that they obey the other laws of dynamics; that they exert forces only at impact and not at a distance; and that they make perfectly elastic collisions (the total kinetic energy before collision is the same after). He then works out these fundamental ideas so as to explain various known gaseous phenomena and to derive new theoretical results, such as his distribution law for molecular velocities. Just before publishing the paper, Maxwell wrote to Stokes in 1859, saying:

> I do not know how far such speculations may be found to agree with facts, ... and at any rate as I found myself able and willing to deduce the laws of motion of systems of particles acting on each other only by impact, I have done so as an exercise in mechanics. Now do you think that there is any so complete a refutation of this theory of gases as would make it absurd to investigate it further so as to found arguments upon measurements of strictly "molecular" quantities before we know whether there be any molecules?[9]

In his paper, Maxwell at one point describes what he is doing as constructing a "physical analogy" between a "system of bodies" (molecules) and gases. But if he really intended this as an analogy, it is very different from the analogy for the electrical field he describes in his earlier paper. The analogy in the present paper is expressed by a set of assumptions about gases, albeit ideal gases, not by assumptions about some distinct system such as his imaginary fluid (as an analogue of the electric field) or a swarm of bees (as an analogue for a collection of gas molecules). The molecules he postulates in his

---

9. Elizabeth Garber, Stephen G. Brush, and C. W. F. Everitt, eds., *Maxwell on Molecules and Gases* (Cambridge, MA: MIT Press, 1986), p. 279.

1860 paper are supposed to be constituents of the gases of which he speaks. Moreover, in the same paragraph in which he uses the term "physical analogy" he also speaks of the assumptions he introduces about the molecules as a "theory" capable of explaining observed phenomena involving gases, unless contradicted by experimental results.

Before I describe in general terms the method Maxwell is using here, why he is using it, and how it is different from any of the others we have considered previously, I want to give a fuller idea of how Maxwell in fact proceeds in the paper. He begins as follows:

> So many of the properties of matter, especially when in the gaseous form, can be deduced from the hypothesis that their minute parts are in rapid motion, the velocity increasing with the temperature, that the precise nature of this motion becomes a subject of rational curiosity. (1:377)

He continues:

> Daniel Bernouilli, Herapath, Joule, Kronig, Clausius, etc. have shown that the relations between pressure, temperature, and density in a perfect gas can be explained by supposing the particles to move with uniform velocity in straight lines, striking against the sides of the containing vessel and thus producing pressure. (1:377)

In addition to deriving the relations he notes, Maxwell also mentions the task of determining the mean length of the path of a molecule between successive collisions ("mean free path") from so-called transport phenomena of gases (internal friction, heat conduction, and diffusion). He then writes:

WHAT TO DO IF YOU CANNOT ESTABLISH A THEORY

In order to lay the foundation of such investigations on strict mechanical principles, I shall demonstrate the laws of motion of an indefinite number of small, hard, and perfectly elastic spheres acting on one another only during impact.... If experiments on gases are inconsistent with the hypothesis of these propositions, then our theory, though consistent with itself, is proved to be incapable of explaining the phenomena of gases. In either case it is necessary to follow out the consequences of the hypothesis. (1:377–78)

One of the important molecular explanations Maxwell offers in the paper is that of the known relation between pressure, volume, and temperature of a gas, given by the so-called ideal gas law. From his dynamical assumptions he derives the following equation relating the pressure p of the particles (molecules) on a unit area of the wall of the container to other molecular quantities:

$$p = 1/3 MNv^2,$$

where M = the mass of each particle, N = the number of particles in a unit of volume of the gas, and v = the mean velocity of the particles. The density ρ (total mass per unit volume) of the particles is MN. So we get the result that p is proportional to ρv^2. Now, if we associate ρ with the density (mass per unit volume) of the gas in the container, and v^2 with a quantity proportional to the absolute temperature of the gas, then we get the ideal gas law in the form that the pressure of the gas is proportional to its absolute temperature and inversely proportional to its volume.

So far it might seem that Maxwell is proceeding just like a hypothetico-deductivist, and unlike a Newtonian or Millian inductivist. Like the hypothetico-deductivist, and unlike the inductivist, he begins, without any experimental proof or evidence, by making a set of theoretical assumptions. And like a hypothetico-deductivist, he proposes to "follow out the consequences" of these assumptions. If the consequences are shown to be incompatible with experiments, then the explanations offered by the theory will have to be rejected. Unlike most hypothetico-deductivists, however, he is silent about the situation in which the consequences are found to be experimentally verified. He does not say, for example, that the fact that he can derive the ideal gas law is confirming evidence for the molecular theory. However, before we conclude that Maxwell's method is a Popperian falsificationist version of hypothetico-deductivism (according to which only falsification, not verification, is possible), more needs to be said about his actual procedure.

The bulk of Maxwell's paper (30 out of 32 pages) consists of what he calls "propositions," which are statements of problems to be solved, using the fundamental assumptions of his theory, mathematics, the solution to previous problems, and new assumptions introduced for the solutions of these problems. (These are the "exercises" he refers to in his letter to Stokes.) One of these, the derivation of the ideal gas law, is noted above. Here I want to consider a different one, his derivation of the so-called Maxwell distribution law for molecular velocities (probably the most important result in the paper). It illustrates aspects of Maxwell's method that make its aims and procedures different from hypothetico-deductivism or the inductivism of Newton and Mill.

The problem is given as follows:[10]

> *Proposition IV*: To find the average number of particles [molecules] whose velocities lie between given limits, after a great number of collisions among a great number of equal particles.

---

10. Readers who want to avoid formulas and get to the bottom line may do so by moving directly to the material following formula (7).

WHAT TO DO IF YOU CANNOT ESTABLISH A THEORY

Maxwell's solution begins like this. Let N be the total number of molecules in a sample of gas. Let x, y, z be the components of velocity for a molecule using rectangular coordinates. Let $Nf(x)dx$ be the number of molecules whose x-component of velocity lies between x and $x + dx$, where $f(x)$ is some function of x to be determined. ($f(x)dx$ can be thought of as the fraction of molecules with x velocities between x and $x + dx$, or as the probability that a molecule in the sample has an x coordinate of velocity in that range.) Similar definitions are given for $Nf(y)dy$ and $Nf(z)dz$.

Maxwell assumes that the velocities x, y, and z are independent, so that the number of molecules with x-components of velocity between x and $x + dx$, and with y-components between y and $y + dy$, and with z-components between z and $z + dz$, is

$$Nf(x)f(y)f(z)\,dxdydz. \tag{1}$$

If we think of $f(x)dx$ as the probability that a molecule will have an x-component of velocity between x and $x + dx$, then what Maxwell is assuming here is that this probability is independent of the probability that the same molecule has a y-component between y and $y + dy$, and similarly for the z-component. Since these probabilities are independent, they can be multiplied to yield the probability that the molecule has x, y, and z components, respectively, between x and $x + dx$, y and $y + dy$, and z and $z + dz$.

Now, we suppose that the N molecules in the sample of gas start from some common origin at the same time. Then from (1), the number of molecules in an element of volume $dx\,dy\,dz$ will be

$$Nf(x)f(y)f(z). \tag{2}$$

And the fraction of the total N in that volume element will be

$$f(x)f(y)f(z). \qquad (3)$$

Maxwell next assumes that since the directions of the coordinates are arbitrary, the number of molecules in the unit volume $dx\,dy\,dz$ depends only on the distance of this volume element from the origin. This is to assume that the fraction of molecules in a unit volume of velocity space[11] is a function of the distance of that volume from the origin (i.e., the fraction does not depend on the direction of molecules in that volume but only on their speeds). But the square of the distance of a point in a space given by coordinates x, y, z, is $x^2 + y^2 + z^2$. So Maxwell is assuming that the fraction of molecules with velocities between x and $x + dx$ and y and $y + dy$ and z and $z + dz$ is some function of $x^2 + y^2 + z^2$; that is,

$$f(x)f(y)f(z) = \Omega(x^2 + y^2 + z^2). \qquad (4)$$

The only mathematical solution possible is an exponential one, $f(x) = Ce^{Ax^2}$, and similarly for y and z, where C and A are to be determined. If A is positive, then the number of molecules given by (2) will increase with velocity toward infinity. Therefore, we make A negative and equal to $-1/\alpha^2$ so that the number of molecules with x-components of velocity between x and $x + dx$ is

$$NCe^{-x^2/\alpha^2}\,dx. \qquad (5)$$

---

11. We can think of a three-dimensional "velocity" space, each coordinate of which represents one of the three velocity coordinates. A point in this space will represent the velocity of some molecule.

If we integrate this from $x = -\infty$ to $x = +\infty$ the result is equal to N, the total number of molecules, since we are integrating over all possible x-components of velocity. The result of this integration is $NC\sqrt{\pi}\alpha$. Equating this to N and solving for C, we get $C = 1/\alpha\sqrt{\pi}$. Therefore, from (5) we derive that the number of molecules whose x-components of velocity lie between x and $x + dx$ is

$$N(1/\alpha\sqrt{\pi})(e^{-x^2/\alpha^2})dx. \qquad (6)$$

There are similar expressions for y- and z-components of velocity.

The number of molecules whose speed (independent of direction) is between v and $v + dv$ can be determined to be

$$N(4/\alpha^{3\sqrt{\pi}})(v^2 e^{-v^2/\alpha^2})dv. \qquad (7)$$

(6) and (7) express Maxwell's formulation of the distribution law for molecular velocities and speeds.

Several points deserve emphasis. The derivation above is an important component of what I call the "theoretical development" of a theory. Maxwell begins his paper with a set of simplifying assumptions about molecules in gases, viz., that they are perfectly elastic spheres subject only to contact forces. The rest of the paper consists in developing these ideas by responding to a series of questions, or rather tasks, including ones expressed as follows:

> *Proposition I*: Two spheres moving in opposite directions with velocities inversely as their masses strike one another; to determine their motions after impact.
>
> *Proposition II*: To find the probability of the direction of the velocity after impact lying between given limits.
>
> *Proposition IV*: (his task for developing the distribution law).

Proposition XII: To find the pressure on unit of area of the side of the vessel due to the impact of the particles upon it (the task leading to the ideal gas law).

Each of these tasks Maxwell completes by a series of mathematical calculations and derivations using the initial simplifying assumptions about gas molecules and their motions. For this purpose he frequently introduces assumptions beyond those made initially. These assumptions are usually not justified empirically (e.g., Maxwell's assumption that the components of velocity of a molecule are independent), but are made just to simplify the calculations. Moreover, frequently the task is completed by arriving at a proposition that is not testable, at least not in Maxwell's day. This is so with his distribution law for molecular velocities. Maxwell did not know how to experimentally determine N (the number of molecules in the sample of gas). Direct experimental tests of this law became possible only in the twentieth century with molecular beam experiments.[12]

This I take to be an important difference between Maxwell and both hypothetico-deductivists and inductivists. His aim in the paper is to work out the theory by answering a series of questions about the entities and processes postulated, whether or not the answers he arrives at are testable, or known to be true, or used to make predictions. Without such "theoretical development" the theory is sketchy,

---

12. In 1860, Maxwell was unable to experimentally establish Avogadro's number. If he had he could have determined N by choosing a quantity of gas equal in mass to the molecular weight of the gas. Beginning in 1865 with the work of Josef Loschmidt and others, including Lord Kelvin, the size of molecules, and of Avogardo's number, could be estimated. For a discussion of this, see Stephen G. Brush, *Statistical Physics and the Atomic Theory of Matter* (Princeton, NJ: Princeton University Press, 1983), p. 57ff. For a description of some initial molecular beam experiments in the 1920s, see Jas. P. Andrews, "The Direct Verification of Maxwell's Law of Molecular Velocities," *Science Progress* 23 (1928–9), 118–23.

minimal, incomplete. To be sure, with hypothetico-deductivists as well as inductivists there is some "working out" of the theory, but just enough to be able to derive consequences (in the "deductive" part) that can be, or have been, tested empirically. What is the point of deriving a molecular distribution law if you can't test it so as to confirm or disconfirm it or the theory from which it was generated (as required by the hypothetico-deductivist); or if you can't establish it by "causal-inductive" reasoning and use it to explain known phenomena (as required by Newton in "analysis" and "synthesis," and by Mill in the three stages of his "deductive method")? What Maxwell wants, as he puts it, is to "lay the foundations" of his investigations into the properties of matter "on strict mechanical principles"—that is, principles, mathematically expressed—that govern bodies in motion subject to forces obeying Newtonian dynamics. In order to do this he wants to say in a mathematically precise way how such bodies move after they collide, how far they move between collisions, how their velocities are distributed, and so forth. Why does he want this, if not simply to establish or at least test the assumptions of the theory?

Here we need to recall Maxwell's aim when you have a set of facts that have been established about a range of phenomena, but no theory about those phenomena that you can establish experimentally. There are three things Maxwell wants in such cases: a physical (and not merely a mathematical) way of understanding the phenomena, one that is expressed precisely using mathematics, and one that is worked out in detail. The latter two are nonepistemic; achieving them does not provide a reason for believing the resulting conception is true. Nor, in general, does achieving the first. As noted earlier in discussing physical analogies, a "way of understanding" observed phenomena by means of an analogy may be developed with mathematical precision without being successful in the ways

Maxwell seeks. (The analogy may not accurately reflect the laws governing the original system; it may be too complex for the mind to grasp, etc.) But even if it is a good one in these respects, that does not entail that inferences can be drawn to the truth or probability of claims about the original system.[13]

Similarly, in the case of an "exercise in mechanics," assumptions may be introduced that are worked out and precise, but not adequately so. (The working out may be extensive but disjointed, not unified; and the precision may contain inconsistencies or be inappropriate in other ways.) But even if the working out and the precision are exemplary, it does not follow from this that the theory is probable or believable. Admittedly, if a theory is precisely formulated and worked out, it may be easier to test than if it isn't, but that doesn't make it (more) probable or believable.[14] In chapter 3, when I discussed "inference to the best explanation," I argued that various criteria included under Lipton's "loveliness" classification are non-epistemic ones. I include the present ones in this class as well.

Maxwell's development of his kinetic-molecular theory does have these virtues: it is expressed precisely using mathematics, and it is worked out in considerable detail. Also, it provides a physical, and not merely a mathematical, way to think about gases. I will assume that it does a good job in these respects, at least by standards

---

13. The exception will be physical analogies in which the analogue is a known system that is very similar to the original one. But Maxwell wants to cast his net very widely to include cases in which such inferences cannot be made from the analogue system—indeed, in which the analogue system is purely imaginary.
14. An example that approximates this situation is one given in chapter 3: Descartes' law of conservation of motion and his three laws of motion. The fact that this is reasonably precise and well worked out (e.g., in Descartes' laws of collision) is not a good reason to believe them. Nor did Descartes take it to be, since he insists on proving his fundamental law of conservation of motion from God's existence and his constancy.

of molecular physics at the time. Does it provide a way of understanding gaseous phenomena, and is this epistemic?

It provides a way of understanding the phenomena in terms of a causal mechanism responsible for them. (By contrast a physical analogy provides a way of understanding the phenomena by conceptualizing those phenomena in a useful way without necessarily indicating a cause or causal mechanism.) But this "way of understanding" itself needs to be understood carefully. What Maxwell is doing in this paper is showing how it is *possible* to understand the behavior of gases by reference to mechanical causes. Whether the particular mechanistic assumptions he makes in the paper are true is not the question at issue. Nor does he seek to establish that some mechanistic assumptions or other about gases are true. His question is whether it is even possible to provide a mechanistic account that could explain known properties of gases (and would satisfy criteria of precision and completeness). Maxwell seeks to propose a possible "way of understanding" gaseous behavior mechanically.[15]

15. By contrast with his 1855 paper, "On Faraday's Lines of Force," in which he employs his method of physical analogy to understand electrical and magnetic phenomena, in a later 1861–62 paper on electricity, entitled "On Physical Lines of Force" (*Scientific Papers*, vol. 1, pp. 451–513), Maxwell uses his "exercise in mechanics" strategy. He shows how a purely mechanical fluid containing rotating vortices to act as idle wheels could produce electromagnetic properties. As in the kinetic theory paper he invokes speculative hypotheses by explicitly identifying the electromagnetic field with the mechanical system he describes. In this 1861–62 paper, he writes: "I propose now to examine magnetic phenomena from a mechanical point of view, and to determine what tensions in, or motions of, a medium are capable of producing the mechanical phenomena observed" (p. 452). In his later 1873 work, *Treatise on Electricity and Magnetism* (New York: Dover, 1954), vol. 2, p. 470, he writes: "The attempt which I then made [in "Physical Lines of Force"] to imagine a working model must be taken for no more than it really is, a demonstration that mechanism may be imagined capable of producing a connexion mechanically equivalent to the actual connexion of parts of the electromagnetic field. The problem of determining the mechanism required to establish a given species of connexion between the motions of the parts of a system always admits of an infinite number of solutions." Nancy Nersessian classifies both of Maxwell's papers on electricity (1855 and 1861–62) as employing the "method of physical analogy." On this point, I take issue with her. See her "Maxwell and 'the Method of Physical Analogy,'" in David B. Malament, ed., *Reading Natural Philosophy* (Chicago: Open Court, 2002), p. 166.

That is why he can introduce various mechanical assumptions for which he gives no empirical arguments, including ones that he takes to be simplifications or idealizations.

By analogy, before the detective attempts to gather evidence establishing that the master of the house was strangled to death by his servant, he may ask how a strangulation was even possible (given the relative sizes and strengths involved, given the suicide note left by the master, etc.). The detective needs to, or at least may want to, establish first that it could have been strangulation by the servant. Whether you are Maxwell or a detective, before trying to establish that a particular theory is true, it may be very important for you to consider how that particular theory, or one of that type, could be true. If you are successful in this endeavor and produce a theory giving a possible mechanism, nothing follows about whether the theory you have constructed, or even the type of theory (e.g., a mechanical one), is in fact true or even probable, no matter how adequately worked out it is.[16] Accordingly, this is not an "epistemic virtue" of the theory developed, in the sense of that expression I have been using (one that provides, or helps to provide, a good reason to believe that the theory is true). At best, we might say, it provides a good reason to believe that the theory, or one of that type, *could* be true.

Why pursue the question of whether a mechanical explanation of gaseous phenomena is possible? Recall Maxwell's first sentence in his paper:

> So many of the properties of matter, especially when in the gaseous form, can be deduced from the hypothesis that their minute

---

16. There is a very good discussion of the importance of this type of strategy in Darwin by Victor Di Fate in his Ph.D. dissertation, *Real Causes for Revision: Scientific Methods in the so-called Vera Causa Tradition*, Johns Hopkins University, 2010.

parts are in rapid motion, the velocity increasing with the temperature, that the precise nature of this motion becomes a subject of rational curiosity.

In a later paper (which I will discuss in the next section), Maxwell writes as follows:

> when a physical phenomenon can be completely described as a change in the configuration and motion of a material system, the dynamical explanation of that phenomenon is said to be complete. We cannot conceive any further explanation to be either necessary, desirable, or possible, for as soon as we know what is meant by the words configuration, motion, mass, and force, we see that the ideas which they represent are so elementary that they cannot be explained by means of anything else. (2:418)

These two quotations suggest how to answer the question at the beginning of the previous paragraph. Dynamical explanations of phenomena are important to Maxwell for methodological reasons expressed in the second quotation, and also for a reason that he mentions later in the same paper: their success in astronomy and electrical science. (Perhaps for Maxwell the latter provides some conditional inductive reason to think that if the postulated particles exist, they too will be subject to dynamical laws.) Hence, his aim is to furnish such a theory for gaseous phenomena. He is further motivated to do so because, as the first quotation suggests, so many of the known properties of gases can be deduced from dynamical assumptions. But the "precise nature" of this motion remains to be worked out. Is that even possible? The main project in Maxwell's paper is to demonstrate a way to do so.

One of the reasons Maxwell cannot claim to have established his theory, despite his success in its theoretical development, is that he offers no epistemic reasons for supposing that spherical molecules exist that interact only at impact and are subject to dynamical assumptions. But another reason is the existence of what in a later paper (to be discussed next) he calls "hitherto unconquered difficulties." The most important of these he considers to be a derivation of the ratio of the specific heat of a gas at constant pressure to its specific heat at constant volume. According to theoretical calculations from his assumptions, in the best case, assuming that molecules are not even spherical but mere material points incapable of rotation, the ratio is 1.66, whereas the observed value is 1.4. This difference Maxwell considers "too great for any real gas." And if we suppose that molecules are spherical and can vibrate, so that there are at least six degrees of freedom, the theoretical calculation of specific heat ratios will be a maximum of 1.33, which is too small for hydrogen, oxygen, nitrogen, and several other gases. Maxwell writes: "This result . . . seems decisive against the unqualified acceptation of the hypothesis that gases are such systems of hard elastic particles." (1:409) In a later paper he considers this "to be the greatest difficulty yet encountered by the molecular theory." (2:433)

Despite this problem he does not reject the program in the paper, because his task there is not to establish the particular mechanism postulated, but to show how a mechanism of that sort could be developed theoretically to account for certain observed phenomena and to yield new results that would provide answers to questions about the mechanism postulated. Given his other successes, including a derivation of various known phenomena and of new untested laws, the fact that he gets a negative result does not deter him from continuing to work out the theory. In an 1866 paper, "The Dynamical Theory of Gases," he changes the force assumption from contact forces only to forces that vary inversely as the fifth power of the distance. And in the 1875 paper, to which I now turn, he makes no assumptions about intermolecular forces or the shape of molecules.

3. A Method of Physical Speculation

In 1875, fifteen years after the publication of his first kinetic theory paper discussed above, Maxwell published a paper in *Nature* with the title "On the Dynamical Evidence of the Molecular Constitution of Bodies."[17] As the title itself suggests, Maxwell here is not treating the idea of the "molecular constitution of bodies" as a mere possibility to be explored by means of an "exercise in mechanics." Rather, it is something for which he can give (at least some) evidence.

Near the beginning of his paper he writes:

> In attempting the extension of dynamical methods to the explanation of chemical phenomena, we have to form an idea of the configuration and motion of a number of material systems, each of which is so small that it cannot be directly observed. We have, in fact, to determine, from the observed external actions of an unseen piece of machinery, its internal construction. (2:419)

Any dynamical assumptions applied to postulated molecules as constituents of gases will be facing the problem that these

[17]. *Scientific Papers*, vol. 2, pp. 418–38.

constituents and their motions are unobservable. Maxwell continues:

> The method which has been for the most part employed in conducting such inquiries is that of forming an hypothesis and calculating what would happen if the hypothesis were true. If these results agree with the actual phenomena, the hypothesis is said to be verified, so long, at least, as some one else does not invent another hypothesis which agrees still better with the phenomena. (2:419)

Maxwell rejects the use of this method of hypothesis on the grounds that its users "are compelled either to leave their ideas vague and therefore useless, or to present them in a form the details of which could be supplied only by the illegitimate use of the imagination." Even if the "results agree with the actual phenomena," Maxwell will not infer that the hypothesis is true or probable on this basis alone, since a conflicting hypothesis may also agree with the phenomena. In this paper, Maxwell is in the following position. He cannot experimentally prove the molecular assumptions he will introduce. But he doesn't want to simply introduce a hypothesis without argument and then derive testable consequences. Instead, he will use a "method of physical speculation" that is weaker than proof but stronger than either the hypothetico-deductive method or an "exercise in mechanics." He writes:

> When examples of this method of physical speculation have been properly set forth and explained, we shall hear fewer complaints of the looseness of the reasoning of men of science, and

the method of inductive philosophy will no longer be derided as mere guess-work. (2:420)

Maxwell does not spell out this method but illustrates its use in his practice. He applies it to a situation in which what is sought is a theory explaining the observed behavior of macrobodies in terms of the behavior of unobservable microbodies of which they are composed. But I think it applies more broadly to any theory, not directly testable, that is designed to explain a range of observable phenomena. The only way to understand what Maxwell has in mind is to see what he actually does in developing his molecular theory using it. The method contains elements of the "exercise in mechanics," but there is a crucial addition.

Independent Warrant. At the beginning of his paper Maxwell gives the two reasons cited in the earlier quotes for using dynamical assumptions in the explanations of observed properties of gases, or more generally in explaining what he calls "chemical phenomena." One is a methodological appeal to the completeness and fundamentality of such explanations. The other is their success in astronomy and electrical science. Now he continues:

> In studying the constitution of bodies we are forced from the very beginning to deal with particles which we cannot observe. For whatever may be our ultimate conclusions as to molecules and atoms, we have experimental proof that bodies may be divided into parts so small that we cannot perceive them. (2:420)

In this 1875 paper, Maxwell does not say what such "experimental proof" is. But he may well be thinking of various claims, made in his book *Theory of Heat* (first published in 1871), starting with the idea that it has been experimentally established that heat is not a substance (caloric) but a form of energy.[18] The energy of a body, he continues in that book, is either kinetic energy due to motion or potential energy due to the body's position with respect to other bodies. But, he claims, heat cannot be the latter, because the presence of another body is not necessary for heat radiation. So it is due to motion, but not that of the body as a whole, since a body radiates heat even when stationary. He concludes: "The motion we call heat must therefore be a motion of parts too small to be observed separately.... We have now arrived at the conception of a body as consisting of a great many small parts, each of which is in motion. We shall call any one of these parts a molecule of the substance."[19]

In the 1875 paper, he writes:

> Hence, if we are careful to remember that the word particle means a small part of a body, and that it does not involve any hypothesis as to the ultimate divisibility of matter, we may consider a body as made up of particles, and we may also assert that in bodies or parts of bodies of measurable dimensions, the number of particles is very great indeed. (2:420)

---

18. By the mid-nineteenth century, the experiments of Rumford and Davy at the end of the eighteenth century, showing that if caloric exists it must be weightless, and showing that mechanical work can produce an indefinite quantity of heat, were considered decisive against the caloric theory. Also, experiments by Joule in the 1840s on heat produced by the friction of bodies established a quantitative relationship between mechanical work and heat. Maxwell is thinking of the latter when, near the beginning of his *Theory of Heat* (9th ed. [New York: Dover, 2001], p. 77, he writes):
> Such evidence [as to the nature of heat] is furnished by experiments on friction, in which mechanical work, instead of being transmitted from one part of a machine to another, is apparently lost, while at the same time, and in the same place, heat is generated, the amount of heat being in an exact proportion to the amount of work lost. We have, therefore, reason to believe that heat is of the same nature as mechanical work, that is, it is one of the forms of Energy.

19. Maxwell, *Theory of Heat*, pp. 304–5.

The idea, then, is that experiments on heat provide some evidence that bodies are composed of unobservable particles, which Maxwell will call molecules, whether such particles are themselves divisible or indivisible.

I call such initial reasons in favor of the theory "independent warrant." The reasons can be of different sorts and may include (a) appeals to experimental results and observations, arrived at independently of the theory in question; these may provide a causal-inductive or an analogical basis for supposing that the observable macrosystem is composed of some type of unobservables that produce some of the observed behavior of the macrosystem; (b) a methodological appeal to the "fundamental" character and simplicity of the principles being applied to those unobservables; and (c) an appeal to the success of these principles in other domains when applied to objects with the same or similar properties as those attributed to the unobservables. The reasons offered may vary in their strength, but they are not of the form "if we make these assumptions then we can explain and predict such and such phenomena," and they are not sufficiently strong to prove that the theory is true. If they include (a) and (c), as Maxwell seems to require, then they provide some epistemic reasons for believing the hypotheses in question that are independent of the explanatory and predictive power of the assumptions. And if they include (b), and are mathematically tractable, they provide some nonepistemic methodological reasons in favor of the theory. What Maxwell seeks as his basic assumptions are ones that have independent warrant. And, in the present case, he writes:

> Of all hypotheses as to the constitution of bodies, that is surely the most warrantable which assumes no more than that they are material systems, and proposes to deduce from the observed

phenomena just as much information about the conditions and connections of the material system as those phenomena can legitimately furnish. (2:420)[20]

Theoretical Development. Maxwell provides a theoretical development of his assumptions that is different from the one in the 1860 paper. In 1860, he introduced the assumption that the molecules are spherical and interact only at impact. Since this lacks epistemic independent warrant (which is fine for the purposes of an "exercise in mechanics"), in the present paper there is no assumption regarding the shape of molecules and there is no restriction of forces between molecules to contact ones. Instead he uses a very general virial equation derived by Clausius from classical mechanics as applied to a system of particles constrained to move in a limited region of space, and whose velocities can fluctuate within certain limits. Maxwell writes the equation as follows:

$$pV = (2/3)T - (2/3)\sum\sum(1/2Rr).$$[21]

---

20. A very similar approach is taken by Maxwell in an 1875 article "Atom" for the *Encyclopedia Britannica*. See *Scientific Papers*, vol. 2, p. 451.
21. The quantity on the left represents the pressure of the gas or fluid multiplied by the volume of its container and can be directly measured. T is the kinetic energy of the total system of particles. R is the force of attraction or repulsion between two particles. And r is the distance between two particles. The quantity $1/2Rr$, Clausius calls the virial of the attraction or repulsion. The sum is double since the virial for each pair of particles must be determined and then the entire sum of these is taken. Clausius's paper was published in German in 1870, with an English translation in *Philosophical Magazine* 40: 122–27. The latter is reprinted in Stephen G. Brush, *Kinetic Theory*, vol. 1 (Oxford, UK: Pergamon, 1965), pp. 172–78. The general theorem yields the result that the mean value of the kinetic energy of such a system of particles equals the mean value of the virial. In the special case of a gas, considered to be composed of such particles, where the gas is acted on by an external pressure p and confined to a volume V, the theorem can be expressed in the form Maxwell gives it above.

In using this equation Maxwell is *assuming* that gases and fluids are composed of unobservable particles that satisfy the equation. This is not an assumption established by proving the equation itself. (In his discussion, Maxwell uses the terms "particle" and "molecule" interchangeably.) Since the equation is derived from classical mechanics, support for which comes from observations of the behavior of observable bodies, he may be supposing that such observations provide some epistemic independent warrant for the conditional claim that if the postulated particles exist, they satisfy the virial equation as well. And, of course, he has some epistemic warrant for supposing that such particles do exist. Neither warrant amounts to proof sufficient to establish the assumption.

The theoretical development proceeds, in part, by taking the virial equation, together with other assumptions that have not been established for molecules, and in some cases, deriving observable conclusions known to be true, and in other cases, deriving conclusions that are untestable. A case of the former is Maxwell's derivation of the ideal gas law, and of known deviations from that law at low temperatures and high densities. Making the assumptions in question, and examining the virial equation above, we may conclude that the pressure of a gas depends on the kinetic energy T of the system of molecules, which is due to the motion of the molecules and to the quantity $(1/2)Rr$, which depends on the forces between them. Maxwell now argues that the pressure of a gas cannot be explained by assuming repulsive forces between molecules. He shows that if it were due to repulsion, then the pressure of a gas with the same density but in different containers would be greater in a larger container than in a smaller one, and greater in the open air than in any container, which is contrary to what is observed. If we suppose that the molecules of the gas do not exert any forces on each other, then the virial equation reduces to $pV = (2/3)T$. Since T is the kinetic energy of the system of particles, and since $T = (1/2)Mc^2$, where c is the mean velocity of a molecule, Maxwell derives the equation $pV = (1/3)Mc^2$. This is the ideal gas law on the assumption that the temperature of a gas is proportional to the mean kinetic energy of the molecules.

Now, continues Maxwell, it is known that real gases deviate from the ideal gas law at low temperatures and high densities. And he asks whether the second factor in the virial equation dealing with forces between molecules, which was ignored in deriving the ideal gas law, can be invoked to explain actual deviations from that law found in experiments. These experiments show that as the density of a gas increases, its pressure is less than that given by the law. Hence, the forces between the molecules must on the whole be attractive rather than repulsive. In the virial equation this is represented by a positive virial. Experiments also show that as the pressure of a gas is increased, it reaches a state in which a very large increase in pressure produces a very small increase in density, so that the forces between molecules are not mainly repulsive.

As an example of an untestable conclusion, if we assume a highly rarefied gas in which the inter-molecular forces can be ignored, then, as above, we get $pV = (1/3)Mc^2$, where c is the mean velocity of a molecule. If we make the idealizing assumption that, in such a rarefied gas the velocities of all the molecules are the same, then from measurements of the quantities p, V, and M, Maxwell determines that the velocity of an oxygen molecule is 461 meters per second, that of nitrogen is 492, and that of hydrogen is 1844, at a temperature of 0 degrees Centigrade. This theoretical result Maxwell had no way of confirming experimentally. Another example of an untestable conclusion in the present paper is Maxwell's distribution law. Here, however, instead of deriving it mathematically as he does in the 1860 paper, he offers a qualitative explanation of its meaning.

"*Hitherto Unconquered Difficulties.*" Another component of his method of physical speculation, which is also present in his "exercise in mechanics," is an enumeration of difficulties. As in 1860, Maxwell mentions the specific heat derivation he invoked in the earlier paper that is inconsistent with observations; he also mentions phenomena that remain to be explained in molecular terms including the transparency of gases and their electrical properties. Maxwell's attitude is this:

> But while I think it is right to point out the hitherto unconquered difficulties of this molecular theory, I must not forget to remind you of the numerous facts which it satisfactorily explains. We have already mentioned the gaseous laws, as they are called, which express the relations between volume, pressure, and temperature, and Gay-Lussac's very important law of equivalent volumes. The explanation of these may be regarded as complete. The law of molecular specific heats is less accurately verified by experiment, and its full explanation depends on a more perfect knowledge of the internal structure of a molecule as we yet possess. But the most important result of these inquiries is a more distinct conception of thermal phenomena. (2:436)

What can one conclude about a theory developed using Maxwell's method of physical speculation? Let us divide the assumptions made by a theory that postulates a realm of unobservables into two sorts: those for which there are epistemic independent warrant arguments, and those for which there are not. In Maxwell's method the most fundamental assumptions of the theory should be of the first sort. If they are, and if the arguments supplied are sufficiently strong, then one can claim to be justified in believing them to be true, even if the assumptions postulate unobservables, and even if the assumptions cannot at the time be proved experimentally.[22] And if a range of observed phenomena is explained by derivation from these assumptions, then justification for the assumptions can be claimed not only on the basis of the epistemic independent warrant but also on the basis of the explained phenomena.[23]

The assumptions for which no epistemic independent warrant is given are ones for which conditional claims are usually made: if we assume such and such, then we can derive the following result. A conditional claim of this sort might have some inductive support (e.g., "if molecules exist, then they satisfy the virial equation—since particles generally do"). Or it might not (e.g., "if molecules are spherical, then they obey the distribution law"). But even if it does have such support, this by itself gives no warrant for the antecedent assumption of the conditional. If the conclusion is determined to be true (e.g., that gases under certain conditions approximately obey the ideal gas law), but there is no epistemic warrant for the assumptions being made to generate that conclusion (e.g., spherical molecules), then since Maxwell explicitly rejects the method of hypothesis, he will not conclude that one is justified in believing the

---

22. Although I have expressed this idea in a manner that suggests "scientific realism," I am not arguing for that position here. An anti-realist could express the same idea by talking about belief in the "empirical adequacy" of the theory. For my own defense of scientific realism, see my "Is there a Valid Experimental Argument for Scientific Realism?," *Journal of Philosophy* 99 (2002): 470–95, reprinted in Achinstein, *Evidence, Explanation, and Realism*.

23. A formal probabilistic representation of this idea is as follows: Let h be a hypothesis or set of hypotheses; let i be the epistemic independent warrant for h; and let e describe a set of known laws and other phenomena derived from h. Then $p(h/i\&e) \geq p(h/i)$. So if h's probability on the independent warrant is high, it will remain at least as high on the additional data e if e is derived from h. Cf. chap. 3, n. 6.

assumptions. What, then, can one conclude about assumptions for which there is no epistemic warrant?

For Maxwell, nothing epistemic. Yet an important way of defending a theory is by showing how it can be developed theoretically. According to Maxwell, this involves formulating assumptions precisely, often mathematically; adding new theoretical assumptions about the unobservables postulated in response to questions about the properties and behavior of those entities; and deriving consequences. Frequently in such a development, new theoretical assumptions are introduced for which no epistemic independent warrant is given, and theoretical consequences are drawn that are not testable. In his earlier 1860 paper, in response to questions he posed regarding molecular velocities, Maxwell developed his theory by adding (undefended) assumptions about the independence of component molecular velocities, leading to a derivation of his (untestable) molecular distribution law. In doing so he did not provide any new or increased epistemic reason to believe his general molecular assumptions or the specific ones needed for the derivation.

Nor is such theoretical development what some have called an "aesthetic" criterion of goodness that adds beauty or simplicity to the theory. (A particular theoretical development may be quite complex and unbeautiful.[24]) In telling us much more about the entities and properties introduced than is done in the central assumptions, its

---

24. This was Duhem's criticism of Kelvin's theoretical development of the wave theory of light in the latter's *Baltimore Lectures* (reprinted in Robert H. Kargon and Peter Achinstein, eds., *Kelvin's Baltimore Lectures and Modern Theoretical Physics* [Cambridge, MA: M.I.T. Press, 1987.]) Kelvin developed the theory by proposing conflicting theoretical models of the ether to interpret various optical phenomena. He proceeded, like Maxwell, by answering a series of questions about the structure of the ether, but the development lacked coherence, since it presented conflicting models for different optical phenomena.

EVIDENCE AND METHOD

main purpose is to add some measure of completeness to the theory by answering a range of questions that might be prompted by considering the fundamental assumptions, and to do so with precision. Completeness and precision are nonepistemic virtues, ones Maxwell seeks in each of his three methods discussed in this chapter. He regards them as valuable for their own sake, and not just for leading to conditional explanations and predictions of phenomena (if they even do so) or for leading to tests of the theory (again if they do so), or just for providing reasons to pursue the theory.[25] Without a theoretical development, he suggests, the basic assumptions are "vague" in the sense of being undeveloped.

Accordingly, while the use of each of Maxwell's three methods is intended to result in theorizing that has nonepistemic virtues, only the "method of physical speculation" aims to result in theories that are defensible both epistemically and nonepistemically. Such theories will have the epistemic virtue that their fundamental assumptions and perhaps others have some causal or inductive independent warrant; and depending on the strength of this warrant and on the known phenomena derived, this may be enough for one to be justified in believing those assumptions.

To be sure, the methods advocated by Newton, Mill, and Whewell are also intended to provide sufficient justification to believe the theory defended by using these methods. But there is an important difference. The methods of Newton, Mill, and

---

25. It would be misleading to say that, for Maxwell, the "theoretical development" constitutes simply what some have called a "logic of pursuit." Maxwell's aim is to employ a method that can be used to show both epistemic and nonepistemic virtues of a theory without proving it. He wants more than simply giving reasons for pursuing the theory or taking it seriously. He wants reasons for believing it to be true, and for concluding that it is a good theory. The "theoretical development" may provide part of one's reason to pursue a theory, but so will the other components of the method. And, as I have emphasized, that is not the raison d'être of this component.

WHAT TO DO IF YOU CANNOT ESTABLISH A THEORY

Whewell are based on the idea that inference to the truth of a scientific proposition or theory requires proof, which these methods are designed by their proponents to enable scientists to provide. Newton and Mill draw a sharp distinction between proof and possibility. Whewell recognizes that there are situations in which you have less than proof (which requires "consilience") but more than mere possibility (in the latter case, for example, when your theory predicts as well as explains phenomena of the same type as those prompting the theory in the first place.) By contrast, although Maxwell grants that proof is always desirable, if and when you can get it, his method of physical speculation is based on the idea that in many situations you have less than proof but more than possibility, or Whewellian success in explaining and predicting phenomena of the sort that prompted the theory. In such situations, depending on the strength of the independent warrant and of the explanations offered, you may be able to infer that your theory (or at least its set of fundamental assumptions) is true, while at the same time recognizing that more theoretical development and experimental support are needed and that unsolved problems remain.

Despite the lack of proof, Maxwell's own belief state with regard to his kinetic-molecular theory was a confident one, which might be characterized as follows:

1. He believed that molecules exist and that the independently epistemically warranted dynamical assumptions about them were true.
2. He believed that he was justified in so believing.
3. He believed that neither he nor anyone else had sufficient experimental evidence to demonstrate that the assumptions he was making in the theory are true.

EVIDENCE AND METHOD

These claims about Maxwell can be supported by examining many of his published and unpublished writings in the 1870s, and not just the 1875 paper in question.[26]

Let's call a belief state of the sort Maxwell was in (one satisfying 1–3 above) a "confident but less than perfect belief state" with respect to a hypothesis h. Now admittedly one can be in such a state without being justified in believing h. But my claim is that one can also be in such a state and be justified in believing that h is true. Suppose I own 85 percent of the tickets in a fair lottery, one ticket of which will be drawn at random, and I believe that I will win because I own 85 percent of the tickets. I am justified in believing this even if I haven't proved or demonstrated that I will win. Or suppose that I am a detective trying to solve a crime, and that I have a good deal of information that suspect number one is the perpetrator: the motive, means, and opportunity all fit, as do the descriptions of some witnesses. On the basis of these facts, I come to believe that this suspect is guilty—even though, let's say, not all the evidence fits exactly, and even though I need more direct and positive evidence for a court of law. In the sort of case I am imagining, I am justified in believing what I do, even if I cannot yet prove it.[27] In relevant respects, in 1875 Maxwell's belief state with regard to molecular theory was analogous to

---

26. Here is a passage from an 1875 article on atoms he wrote for the *Encyclopedia Britannica*: "Having thus justified the hypothesis that a gas consists of molecules in motion, which act on each other only when they come very close together during an encounter, but which, during the intervals between their encounters which constitute the greater part of their existence, are describing free paths, and are not acted on by any molecular force, we proceed to investigate such a system." This, of course, contrasts with Maxwell's much more agnostic epistemic state in 1860, when he wrote his first kinetic theory paper as an "exercise in mechanics."
27. In the lottery case as well as in the detective one, although I believe the hypotheses in question, and I am justified in so believing, I do not know that these hypotheses are true. If so, then, this is another reason (in addition to Gettier counterexamples) to reject the "justified true belief" account of knowledge

these. Employing his "method of physical speculation" the way he did, Maxwell was able to develop a theory with respect to whose basic assumptions he was justifiably in a "confident but less than perfect belief state."

4. Philosophical Implications of Maxwell's Methods

Perhaps under the influence of Descartes and Newton, philosophers of science, in discussing scientific methods, have concentrated on the epistemic aim of proving or establishing theories by evidence, and on the question of what rules to follow to achieve this aim. For Descartes, the "evidence" consists of what he calls "intuitions" and "deductions," which infallibly establish a theory. For Newton, the evidence consists of his "Phenomena," which are to be used to empirically demonstrate the truth of a theory in accordance with his methodological rules. These scientists craved certainty—a priori certainty for Descartes, empirical for Newton. Later philosophers of science weakened the requirement from certainty to probability, or "evidence in favor," of the theory. But, as Maxwell emphasizes, there are situations in which no theory is available to prove or give evidence for, or even if one is available there is no evidence at all or none that will prove it. Maxwell rejected silence as the only strategy in such cases. In any of these situations, there are perfectly legitimate scientific aims, and strategies you can follow to achieve those aims. These strategies are not designed to yield proof of a theory in the case of the method of physical speculation, or evidence in favor of the theory in the case of an "exercise in mechanics," or even a theory in the case of the method of physical analogy. All of them are intended to produce, in different ways, a physical understanding of a set of observed phenomena, the first criterion Maxwell sets for good scientific theorizing. And all of them are intended to satisfy the two other criteria Maxwell requires for good scientific theorizing: being expressed precisely using mathematics, and being worked out in some detail.

To be sure, Maxwell, like any scientist, wanted proof when he could get it. But you can't always get it even when you want it. He writes:

> To conduct the operations of science in a perfectly legitimate manner, by means of methodized experiment and strict demonstration, requires a strategic skill which we must not look for, even among those to whom science is most indebted for original observations and fertile suggestions. It does not detract from the merit of the pioneers of science that their advances, being made on unknown ground, are often cut off, for a time, from that system of communications with an established base of operations, which is the only security for any permanent extension of science. (2:420)

There are variations and modifications of each of Maxwell's three methods that are frequently employed for different purposes and audiences. Maxwell himself uses the bee analogy in his kinetic theory without spelling out in any detail the motions of bees or developing it in mathematical terms. His purpose in the article is mainly a pedagogical one for a general audience. Newton, in Queries 28 and 29 in his *Opticks*, is, I think, using a modified version of Maxwell's method of physical speculation. Query 29 asks the question "Are not rays of light very small bodies emitted from shining substances?" But Newton is not content simply to ask the question, or to do so in a way that merely suggests the answer "of course they are." He offers some "independent warrant" for this answer in Query 28 by giving numerous empirical objections to the rival wave theory defended by his critics Hooke and Huygens. And he offers some "theoretical development" in Query 29 itself by showing how the particle theory can explain various known optical phenomena. The theory is not worked out mathematically with the sort of detail that Maxwell had in mind (though it was later in the eighteenth and early nineteenth centuries). Nor does Newton present sufficient "independent warrant" to make one justified in believing it. But he is doing more than simply proposing a hypothesis for the reader's consideration. He is employing a version, albeit a fairly weak one, of Maxwell's method of physical speculation. Newton here is nicely illustrating an idea that he seems to be rejecting in his *Principia* where he writes that hypotheses (propositions not "deduced from the phenomena" in accordance with his rules) have no place in experimental philosophy. He is introducing the particle hypothesis, which he cannot prove, and he is defending it in a way that conforms to some extent with Maxwell's method.

Maxwell's three methods, even when successfully applied to a set of phenomena so as to produce a type of understanding of them, will not suffice to show that those phenomena constitute evidence that establishes some theory, hypothesis, or law.[28] But that is no weakness in Maxwell's methods. For they are designed for situations

---

28. Of the methods discussed in this book, only Newton's rules will do that. More precisely, of the methods considered, only Newton's rules, if applied in an empirically defensible way to a set of phenomena, will allow one to be justified in believing that the law arrived at using that method is true, because the phenomena constitute strong evidence that the law is true.

in which you have no theory, or you have one for which you have no evidence, or one which the evidence you have is insufficient to establish. These are situations in which you can still engage in scientific "theorizing" and do so properly in ways described by these methods.

INDEX

Analysis
  see Newton
Andrews, Jas. P., 150n
A priori assumption, 13, 17–23

Bayesians, 6n, 7n
Brush, Stephen, 19n, 150n

Carnap, Rudolf, 4n, 10n, 21, 25, 40
Cartwright, Nancy, 36n
Clausius, Rudolf, 162
Cohen, I.B., 43n, 48, 50, 91n
Consilience of inductions,
  see Whewell

Deductive method
  see Mill
Descartes, René, 72–73, 108–110
  Rule 3, 72
Di Fate, Victor, 63, 78n, 80n, 154n
Ducheyne, Steffen, 48n
Duhem, Pierre, 106n, 167n

Eliminative reasoning, 90–92
Empirical principle of objective evidence,
  25, 38
Empirically complete, 19

Epistemic vs. pragmatic justification of rules,
  77–80
E-S evidence
  see Evidence
Evidence
  E-S evidence, 33
  Hempel's satisfaction theory,
    11–12
  Hypothetico-deductive (h-d) view,
    5–6, 11
  and Newton's rules, 120–124
  Objective vs. subjective, 5–10, 33
  Positive relevance view, 10–11
  Potential evidence, 31
  Strong veridical evidence, 31–32
  Subjective evidence, 33
  Three principles of, 23–25
  Veridical evidence, 31–32
Exercise in mechanics
  see Maxwell
Explanation, 35–36
Explanatory connection, 28–30

Feyerabend, Paul, 62
Fine, Arthur, 104n

Galileo, 17–19

INDEX

Harman, Gilbert, 92n, 103n
Harper, William, 45n, 51, 53n, 61n, 70n, 86
Hempel, Carl G., ix–xii, xiv, 4n, 11–12
Hume, David, 30
  see justification of induction
Hypothetico-deductivism, 11, 74
  Newton's rejection of, 74–75

Independent warrant
  see Maxwell
Induction
  Mill on, 87–89
  Newton on, 47–52
Inference to the best explanation (IBE)
  see Lipton, Maxwell
  Combined with Newton's rules, 114
  Five arguments for IBE, 104–107
  Strategy rule, 113

Justification of induction, 116–119

Kelvin, Lord (William Thompson), 167n
Kitcher, Philip, 39–40

Lange, Marc, 49n, 79n, 95n
Laudan, Larry, 106
Lipton, Peter, 34, 94–103
  Lipton's rule, 96
  Loveliness, 95
Loveliness
  see Lipton

McAllister, James, 106n
McGuire, J.E., 46n
McMullen, Ernan, 49n
Mandelbaum, Maurice, 49, 50
Maxwell, James Clerk, 110, 129–130
  Distribution law, 146–149
  Exercise in mechanics, 142–157
  "Hitherto unconquered difficulties," 165
  Independent warrant, 159–162
  Method of physical analogy, 130–142
  Method of physical speculation, 157–171
  Theoretical development, 162–165
  Three requirements for theorizing, 130, 151

Methods
  Analysis, see Newton
  Deductive, see Mill
  Exercise in mechanics, see Maxwell
  Hypothetico-deductive, see
    hypothetico-deductivism
  Inference to the best explanation, see
    Lipton, Whewell
  Physical analogy, see Maxwell
  Physical speculation, see Maxwell
  Synthesis, see Newton
Mill, John Stuart, 48, 49, 75n
  Deductive method, 87–89
Morgan, Gregory, 36n, 107

Nagel, Thomas, 105n
Nersessian, Nancy, 153n
Newton, Isaac, 3, 171–173
  Analysis, 84–85
  Argument for the law of gravity, 52–58
  Extensions of Newton's method, 84–92
  "Hypotheses," 52, 75
  Newton's strategy, 58–65
  "Phenomena," 46–47, 53–54
  Rules, see Newton's rules for the study of
    natural philosophy
  Synthesis, 84–85
Newton's rules for the study of natural
  philosophy, 43–52
  An interpretation, 66–71
  Response to criticisms of the rules, 80–83
  Their value, 72–80
Norton, John, 63, 78n, 117n

Penrose, Roger, 106n
Phenomena
  see Newton
Physical analogy
  see Maxwell
Physical speculation
  see Maxwell
Popper, Karl, 86, 117
Positive relevance, 10–11
Potential evidence
  see evidence
Principles of reasonable belief, 23–25, 36–38

176

INDEX

Probability, 34–35
Problem of evidence, 4–5, 38–41, 120–124

Reid, Thomas, 44n

Satisfaction theory of evidence
   see Evidence
Smith, George E., 51, 60n, 86
Spencer, Quayshawn, 60
Strong veridical evidence
   see Evidence
Subjective evidence
   see Evidence
Synthesis
   see Newton

Theoretical development
   see Maxwell
Threshold concept, 16, 25–26

van Fraassen, 114n, 127–128
Veridical evidence
   see Evidence

Weakness assumption, 10, 13–17
Whewell, William, 44n, 59–63, 78n, 134–136
   Coherence, 93–94
   Colligation, 91–92
   Consilience, 93
   Whewell's rule, 94
Whitman, Anne, 3n

CPSIA information can be obtained
at www.ICGtesting.com
Printed in the USA
BVHW07*0712280718
522858BV00003B/5/P